S0-CRV-010

PITUITARY-ADRENAL FUNCTION

A symposium organized by the Section on Medical Sciences of the A.A.A.S. and presented at the New York meeting on December 28-29, 1949

Organized by

GORDON K. MOE

Secretary of the Subsection on Medicine

Edited by

RUTH C. CHRISTMAN

A Publication of the

American Association for the Advancement of Science

1515 Massachusetts Avenue, N.W., Washington 5, D. C.

1950

Reprinted June, 1951

Printed by
THE HORN-SHAFER COMPANY
BALTIMORE 2, MD.

FOREWORD

Since 1937 the programs of the Subsection on Medicine of the American Association for the Advancement of Science have consisted of invited papers on various aspects of medical science. The advantage of the symposium plan has been apparent from the popularity of the programs and from the wide sales of the several symposium volumes which have been published by the Association. By arranging the program of the section around a central theme it has been possible to capture and hold the interest of relatively large audiences at several sessions over a one- or two-day period.

In preparation for the 1949 program, a list of some twenty suggested symposium topics was sent to more than a hundred distinguished investigators in all fields of medical science. The intense interest in the adrenal cortex generated by the spectacular clinical successes with cortisone and ACTH was reflected in the votes of the specialists whose advice was solicited, the overwhelming majority of whom chose the adrenal as the symposium subject.

The committee proceeded to prepare a program with some misgivings, fearing that the prominent investigators in the field would, by December of 1949, have been talked to death and would probably prefer to remain in their laboratories grinding out new data rather than come to New York to mull over the old. However, one remarkable characteristic of this productive group of scientists became evident when they were invited to contribute to yet another program: they are, without exception, just as enthusiastic about getting together to talk about their work as they are about doing it; and that they have been able to continue their investigations despite such interruption is evident from the speed with which new data appear from their laboratories.

While this collection of papers cannot be expected to include everything known about the chemistry, physiology, and clinical actions of the pituitary-adrenal hormones, it does represent an impressive segment of the work from a score of laboratories whose continuing contributions to this field have advanced it so rapidly.

The publications committee regrets that Dr. J. S. L. Browne and Dr. Abraham White were unable to submit manuscripts for inclusion in the volume. The committee wishes to express its appreciation to those who took part in the symposium and to Dr. Perera who prepared on short notice the introductory chapter.

G. K. Moe,
Secretary, Section Nm

CONTRIBUTORS

FULLER ALBRIGHT, M.D.
Physician, Massachusetts General Hospital; Associate Professor of Medicine, Harvard Medical School, Boston, Massachusetts

CARL T. ANDERSON, A.B.
Institute for Metabolic Research, Highland Alameda County Hospital, Oakland, California

BURTON L. BAKER, PH. D.
Associate Professor, Department of Anatomy, Medical School, University of Michigan, Ann Arbor, Michigan

FREDERIC C. BARTTER, M.D.
Department of Medicine, Harvard Medical School; Medical Service of the Massachusetts General Hospital, Boston, Massachusetts; Surgeon, U.S.P.H.S.; National Heart Institute

JAMES H. BIRNIE, PH. D.
Department of Zoology, Syracuse University, Syracuse, New York

JANE BLOOD, B.S.
Research Assistant, Medical School, University of Michigan, Ann Arbor, Michigan

W. R. BOSS, PH. D.
Department of Zoology, Syracuse University, Syracuse, New York

EVELYN L. CARROLL, A.B.
Massachusetts General Hospital, Boston, Massachusetts

JEROME W. CONN, M.D.
Professor of Internal Medicine, Medical School, University of Michigan, Ann Arbor, Michigan

A. C. CORCORAN, M.D.
Assistant Director, Research Division, Cleveland Clinic Foundation, Cleveland, Ohio

HELEN WENDLER DEANE, PH. D.
Associate in Anatomy, Department of Anatomy, Harvard Medical School, Boston, Massachusetts

ELEANOR DEMPSEY, M.A.
Massachusetts General Hospital, Boston, Massachusetts

KONRAD DOBRINER, M.D.
The Sloan-Kettering Institute, Memorial Hospital, New York City, New York

THOMAS F. DOUGHERTY, PH. D.
Professor and Chairman, Department of Anatomy, College of Medicine, University of Utah, Salt Lake City, Utah

B. EKMAN, M.D.
The Sloan-Kettering Institute, Memorial Hospital, New York City, New York; Visiting Research Associate from University of Lund, Sweden

L. P. ELIEL, M.D.
Division of Clinical Investigation, The Sloan-Kettering Institute, Memorial Hospital, New York City, New York

FRANK L. ENGEL, M.D.
Departments of Medicine and Physiology, School of Medicine, Duke University, Durham, North Carolina

W. J. EVERSOLE, PH. D.
Department of Zoology, Syracuse University, Syracuse, New York

STEFAN S. FAJANS, M.D.
Research Fellow in Medicine of the American College of Physicians, 1949-1950, Department of Internal Medicine, Medical School, University of Michigan, Ann Arbor, Michigan

PAUL FOURMAN, M.D.
Oxford University, England; Rockefeller Traveling Fellow at the Massachusetts General Hospital, Boston, Massachusetts

ROBERT GAUNT, PH. D.
Department of Zoology, Syracuse University, Syracuse, New York

EDGAR S. GORDON, M.D.
Associate Professor of Medicine, Department of Medicine, University of Wisconsin, Madison, Wisconsin

ROY O. GREEP, PH. D.
Harvard School of Dental Medicine, Boston, Massachusetts

ROBERT N. HEDGES, M.D.
Lieut. (j.g.) MC, USNR, U.S. Naval Hospital, Oakland, California

HUDSON HOAGLAND, PH. D.
Executive Director, Worcester Foundation for Experimental Biology, Shrewsbury, Massachusetts; Neurophysiologist, Worcester State Hospital; Research Professor of Physiology, Department of Physiology, Tufts College Medical School

MAXINE E. HUTCHIN, A.B.
Metabolic Research Unit, University of California-U.S. Naval Hospital, Oakland, California

DWIGHT J. INGLE, PH. D.
Research Department, The Upjohn Company, Kalamazoo, Michigan

W. MCK. JEFFERIES, M.D.
Massachusetts General Hospital, Boston, Massachusetts

BETTY JOHNSON, B.S.
Dietician and Research Assistant, Medical School, University of Michigan, Ann Arbor, Michigan

LAURANCE W. KINSELL, M.D.
Director, Institute for Metabolic Research, Highland Alameda County Hospital, Oakland, California

JUDITH LANGE, A.B.
Institute for Metabolic Research, Highland Alameda County Hospital, Oakland, California

CHOH HAO LI, PH. D.
Department of Biochemistry, University of California, Berkeley, California

S. LIEBERMAN, PH. D.
The Sloan-Kettering Institute, Memorial Hospital, New York City, New York

C. N. H. LONG, SC. D., M.D.
Department of Physiological Chemistry, School of Medicine, Yale University, New Haven, Connecticut

LAWRENCE H. LOUIS, SC. D.
Assistant Professor of Biochemistry, Medical School, University of Michigan, Ann Arbor, Michigan

SHELDON MARGEN, M.D.
Schering Research Fellow in Endocrinology, 1948-1949; Damon Runyon Fellow, 1949-1950; Institute for Metabolic Research, Highland Alameda County Hospital, Oakland, California

HAROLD L. MASON, PH. D.
Division of Biochemistry, Mayo Foundation, Rochester, Minnesota

GEORGE D. MICHAELS, PH. D.
Institute for Metabolic Research, Highland Alameda County Hospital, Oakland, California

GORDON K. MOE, M.D., PH. D.
Department of Physiology, State University of New York Medical Center, Syracuse, New York

C. M. OSBORN, PH. D.
Department of Zoology, Syracuse University, Syracuse, New York

O. H. PEARSON, M.D.
Division of Clinical Investigation, The Sloan-Kettering Institute, Memorial Hospital, New York City, New York

GEORGE A. PERERA, M.D.
Associate Professor of Medicine, College of Physicians and Surgeons, Columbia University, New York City, New York

MARSCHELLE H. POWER, PH. D.
Division of Biochemistry, Mayo Clinic, Rochester, Minnesota

C. P. RHOADS, M.D.
The Sloan-Kettering Institute, Memorial Hospital, New York City, New York

FLOYD R. SKELTON, M.D.
Institute of Experimental Medicine and Surgery, University of Montreal, Montreal, Canada

Randall G. Sprague, M.D.
Division of Medicine, Mayo Clinic, Rochester, Minnesota

Betty Sprunger
Research Assistant, Medical School, University of Michigan, Ann Arbor, Michigan

Eleanor H. Venning, Ph. D.
Associate Professor of Medicine, McGill University Clinic, Royal Victoria Hospital, Montreal, Canada

William C. Vogel, B.S.
Great Falls Clinic, Great Falls, Montana

F. C. White, M.D.
Division of Clinical Investigation, The Sloan-Kettering Institute and the Department of Medicine, Memorial Hospital, New York City, New York

H. Wilson, Ph. D.
The Sloan-Kettering Institute, Memorial Hospital, New York City, New York

George W. Woolley, Ph. D.
Former Assistant Director, now Visiting Research Associate, Roscoe B. Jackson Memorial Laboratory, Bar Harbor, Maine; Head, Division of Steroid Biology, The Sloan-Kettering Institute, Memorial Hospital, New York City, New York

CONTENTS

"ADRENOLESCENCE"

GEORGE A. PERERA
College of Physicians and Surgeons, Columbia University, New York

From an ill-defined and subordinate offspring of a parental pituitary, the adrenal cortex now emerges into a position of dominance in the family scene of endocrinological processes. Previously recognized only in terms of its presence or absence as a single member of the hormonal group, its growing pains are over. However, the adrenal cortex remains an adolescent, replete with behavior problems, and one can only speculate on its ultimate degree of maturation and independence through rare glimpses of seeming adult performance.

Will the adrenal cortex ever be an integral and self-sufficient entity? It must be recalled that lower forms of life exist without endocrine glands and perform all the necessary details of growth, metabolism, and reproduction. Dr. C. N. H. Long has made the pertinent comment that "the hormones do not initiate new patterns of cellular function . . . these are an inherent birthright of the cells themselves. All that any hormone does is either to facilitate or inhibit certain types of chemical transformations within the cells." Dr. R. F. Loeb has stated that ". . . perhaps it is well to look on the endocrine system as a very important superstructure which is introduced as an expediting system . . . in (the) complicated organization of society of cells such as exists in higher animals." The concept that the adrenal cortex fulfils a secondary rather than a primary mission seems logical, but it is sometimes forgotten in the enthusiasm produced by its dramatic participation in multiple metabolic activities.

Some thirty steroids have been obtained from adrenal cortical material and close to fifty have been recovered in the urine. Undoubtedly these represent altered and degraded products to a large extent. Details of production and the form or forms in which the actual materials are discharged remain unknown. Certainly the high content of cholesterol and ascorbic acid in the adrenal cortex cannot be ignored and values of these substances are modified conspicuously in relation to the glands' activity. Work is in progress employing isotopic cholesterol and labeled steroids. Quantitative methods of assay, such as by solvent partition and chromatography, are replacing the more crude methods of bioassay. The perfusion of the isolated adrenal with various substances, as undertaken by Pincus and his associates, has already given rise to valuable information. In interpreting new data, the possibility of species differences must be kept in mind, for it has been demonstrated that steroidal patterns are at least quantitatively altered from one animal type to another.

One group of investigators has conceived of a single mother-substance producing both 11-oxygenated and 17-ketosteroids. Others look upon the adrenal cortex as elaborating but one "normalizing" hormone, capable of swinging the metabolic and electrolytic pendulum in either direction, depending upon the state of the body and its deviation from normal under conditions of stress or disease. In opposition to the unitarian view of hormonal production, certain observations must be recalled. Although desoxycorticosterone is not made in the adrenal cortex with certainty, its potency might indicate that it or its equivalent is discharged in too small amounts to be detected by current means of analysis. Primary adrenal cortical insufficiency (Addison's disease) differs in many respects from secondary insufficiency (hypopituitarism) as, for example, in the better maintenance of plasma volume and the larger doses of desoxycorticosterone required to modify serum sodium values in the latter condition. Furthermore, the administration to man of large amounts of cortisone over a long period may have practically no effect on salt and water metabolism; may produce different responses in the patient with intact as compared to diseased adrenals; and has not been found as yet to constitute complete replacement therapy. These and other arguments are not conclusive, but at the moment they influence this author to retain a dualistic concept of desoxycorticosterone-like and cortisone or compound F-like hormones.

Already it is appreciated that behavior of the adrenal cortex is governed in part by that of the rest of the family. It is stimulated from an apparent quiescent state to one of hyperkinetic activity through the adrenocorticotropic secretion of the hypophysis and perhaps independently by epinephrine. Electrolytes may modify its conduct slightly. The adrenal, in turn, influences its pituitary guardian by the blood level of cortical hormones which it releases into the circulation. The exact mechanism of these and other interrelationships remains to be elucidated fully.

It is clear that the steroids elaborated by the adrenal cortex exert profound effects on many aspects of fluid, electrolyte, and even cellular metabolism. In what manner is this achieved? In part through modification of renal tubular function, but this does not tell us the details of active secretory, excretory, and reabsorptive processes and their regulation. Adrenalectomy results in increased carbohydrate utilization, quite possibly a reduction in gluconeogenesis, but this does not delineate completely the mechanism of action or explain such phenonema as the increased sensitivity to insulin under these conditions. Nitrogen metabolism can be affected by the administration of some adrenal cortical hormones, presumably through acceleration of catabolism or decreased anabolism or both, but the final answer still rests upon unknown intracellular and enzymatic processes. The appearance following adrenalectomy of pigmentation, decreased ability to withstand environmental stress, altered capillary permeability, hemodynamic changes, and lymphoid tissue and leucocytic differences are matters still beyond complete understanding. The "unfinished business" of Addison's disease includes an appreciation of the precise mode of hormonal function and interplay. Furthermore, the treatment of adrenal cortical insufficiency should be a simple problem of replacement, but this has not yet been achieved. Lest we become too complacent, it was pointed out by Pfiffner and by Wells and Kendall that only a fraction of the potency of adrenal extracts can be attributed to the known crystalline compounds extracted therefrom. What comprises the rest?

Considerable attention has been focused on the relationship of the adrenal cortex to salt and water metabolism. In order to illustrate the magnitude and complexity of some of the remaining problems, let us limit our examination to the sodium ion. It is accepted that cortical hormones cause sodium retention, and that a liberal salt diet may maintain the adrenalectomized animal in electrolyte equilibrium. Yet it has been suggested that sodium retention occurs when high serum sodium values are produced in adrenalectomized animals, and that the adrenocorticotropic hormone stimulates the adrenal to "normalize" the plasma sodium regardless of the direction of the disturbance. We have recently shown that the Addisonian patient cannot be maintained on even large doses of desoxycorticosterone acetate in the absence of at least some dietary sodium. Although changes in adrenal histology and minor modifications of nitrogen balance have been described by some workers, the addition or subtraction of salt from the diet has not affected the excretion of "corticoids," 17-ketosteroids, or other so-called indices of adrenal cortical activity. There is at yet no adequate explanation for these and other data. Some postulate an adrenal salt hormone influenced only by adrenocorticotropic action; others look upon an independent salt hormone responding to alterations in sodium need through an as yet unexplained pathway; still others feel that a renal mechanism governs electrolyte excretion provided that the kidney is supplied by at least some adrenal stimulus.

A final word about products of the adrenal cortex in the management of disease. The list of responsive disorders, since the original report by Hench and his co-workers of marked improvement in rheumatoid arthritis and related illnesses, now includes a host of conditions in the mesenchymal, metabolic, neoplastic, allergic, and even infectious category. It is apparent that relief requires sustained administration of cortisone or the adrenocorticotropic hormone and that relapses generally occur promptly following the cessation of treatment. It is noteworthy that the vast majority of responsive disorders are in the nature of chronic, often progressive illnesses, the remainder being disturbances of a cyclic or repetitive pattern; there is no convincing proof as yet that the adrenocorticotropic hormone or cortisone elicits more than a remission in the former group, or interrupts more than a single attack in the latter group. It is obvious that the responsive conditions do not represent endocrinological deficiency states, but that results are achieved through the production of at least some measure of hyperadrenocorticalism. The administration of these agents gives rise to numerous hormonal, psychological, metabolic, and other effects, some of which are distinctly dangerous to the patient. The future of these substances remains uncertain, but one may feel reasonably certain that the mechanism of their action will prove a greater contribution of far more significance than is implied in their present dramatic therapeutic role.

The adrenal cortex has already passed through the critical years of discovery and early development. Its fundamental behavior patterns are becoming clarified. Considerable credit must be given for the magnitude and extent of the advances already made. To venture one overall prediction, it is that our adolescent will never mature but will always be a mimic or nonconformist. A superstructure, facilitating what other structures can manage independently, the adrenal cortex will teach us many hidden metabolic secrets through its ability to exaggerate normal regulatory activities in health, during stress and during disease. Future emphasis must be placed primarily upon further elucidation at a cellular level of mechanisms brought to light in hypo- and hyperadrenocortical states.

RECENT KNOWLEDGE ON THE NATURE OF HYPOPHYSEAL ADRENOCORTICOTROPIC HORMONE (ACTH)

CHOH HAO LI
University of California, Berkeley, and University of Upsala, Upsala, Sweden

The physiological and chemical nature of adrenocorticotropic hormone (ACTH) has been reviewed by the writer (Li, 1948). This paper concerns therefore a discussion of the data recently obtained.

ACTH Protein

Adrenocorticotropic hormone (ACTH) isolated from sheep pituitaries possesses some unusual properties and can easily be differentiated from other known protein hormones of anterior hypophyses (Li, 1947). One of the characteristics of ACTH is its resistance to heat treatment. A neutral solution of the hormone retains its biological potency after being kept at boiling water temperature for 120 minutes or longer. Even more remarkable is the fact that the ACTH protein remained active after certain pepsin or acid digestions. It is now clear that the hormonal activity resides in the hydrolyzed fragments (peptide residues) of the original protein molecule (Li, 1948; 1949).

TABLE 1

Certain Physicochemical Data of ACTH

Molecular weight	20,000
Isoelectric point, pH	4.7
Diffusion constant, $D_{20} \times 10^7$	10.4
Sedimentation constant, S_{20}	2.08
S, %	2.30
N, %	15.65
Cystine, %	7.19
Methionine, %	1.93
Essential groups	Amino and phenolic

Physicochemical Data. As summarized in Table 1, the ACTH protein isolated from sheep glands has a molecular weight of 20,000 and an isoelectric point at pH 4.70. It contains 2.3% sulfur which is accounted for by 7.19% cystine and 1.93% methionine. A complete HCl hydrolysate of the hormone has revealed the following amino acids on two-dimensional paper partition chromatography (Consden *et al.*, 1944) using lutidine/H_2O and phenol/H_2O as the solvents: aspartic acid, lysine, serine, glycine, glutamic acid, arginine, threonine, tyrosine, alanine, valine, histidine, proline, phenylalanine, leucines, and tryptophane. No unknown ninhydrin colored spots occurred in the paper.

The Terminal Group. Sanger (1945) has developed a method for the identification and estimation of the free amino groups of protein using the reagent 1,2,4-fluorodinitrobenzene. The terminal residues of insulin, for instance, have been determined by this method and are glycine and phenylalanine.

As further shown by Sanger, the number of terminal groups in a protein will be equal to the total number of open polypeptide chains. Using similar techniques, Li and Porter (1949) have shown that the terminal group in ACTH is alanine and that there is only one mole of alanine per mole of the hormonal molecule. It appears that ACTH protein consists of one open polypeptide chain with an alanine residue as the end group. It is of interest to note that the hormone contains 6 cystine residues per molecule. If the protein has only one single polypeptide chain, the six -S-S-bridges must hold the chain as a coil. Preliminary studies on the effect of reduction of -S-S-groups with cysteine suggest that these groups are not essential for the biological activity.

Essential Groups. In an earlier paper, (Li *et al.*, 1946) we have shown that the free amino and tyrosine groups are essential for the adrenocorticotropic activity of ACTH. The essentiality of the free amino group is now confirmed by a more specific reaction using acetic anhydride as the acetylating agent.

TABLE 2

Effect of Acetylation on Adrenocorticotropic Hormone

Preparation	Daily* Dose (mg)	No. of Rats	Body Weight (g)	Adrenals (mg)
Untreated	0.2	24	128.5 (104-155)	22.2 (15-29)
Treated	0.2	11	136.8 (102-159)	18.6 (12-23)

* Bioassays were carried out in hypophysectomized male rats operated at 40 days of age; injections were instituted on the day of hypophysectomy for 15 days. The average adrenal weight of the controls was 12.0 mg.

Three hundred milligrams of ACTH was dissolved in 10 cc half saturated sodium acetate and kept at 0°C, while 0.4 cc acetic anhydride was added dropwise with constant stirring. The reaction was continued for 30 minutes. The solution was then dialyzed and the salt-free solution was recovered by lyophilization. The free amino nitrogen of the acetylated sample was determined by the Van Slyke (1929) nitrous acid method and the acetyl content by the procedure essentially the same as that described by Herriott (1935). It was found that a decrease of 2.7×10^{-4} mM amino nitrogen per milligram of the acetylated hormone has occurred with an accompanied increase of 2.8×10^{-4} mM acetyl concentration. Thus, it may be concluded that acetic anhydride reacts only with the free amino group of the ACTH molecule under the conditions described.

When the acetylated hormone was assayed in hypophysectomized male rats by the maintenance test (Li and Evans, 1947) a definite lowering of the adrenocorticotropic activity was observed (Table 2). From these data it may be said that the activity of ACTH depends upon the intactness of the free amino groups.

Effect on the Glycogenesis in Isolated Diaphragm. The influence of ACTH on carbohydrate metabolism is well known. It produces glycosuria in normal rats force-fed with high carbohydrate diet and enhances diabetogenecity in alloxan-treated animals (Li and Evans, 1947). Recent experiments with *normal* rat diaphragms (Li *et al.*, 1949) indicate that ACTH, indeed, inhibits the insulin effect in promoting glycogen storage. There is no change in the glucose uptake of the diaphragm of rats injected with ACTH, but glycogenolysis occurs in these diaphragms.

TABLE 3

Glucose Uptake and Glycogenesis by Diaphragms of Normal and Hypophysectomized Rats Treated with ACTH

Experiment	Glucose Uptake			Glycogenesis	
	Without insulin	With insulin	Initial Glycogen	Without insulin	With insulin
			Normal Rats		
Control	2.23±0.13* (20)†	3.78±0.37 (10)	3.95±0.36 (20)	0.22±0.09 (20)	1.52±0.21 (10)
ACTH	2.43±0.14 (6)	3.61±0.10 (12)	3.55±0.26 (18) $p=0.3$‡	−0.19±0.19 (6) $p=0.05$	0.54±0.24 (12) $p<0.01$
ACE	2.15±0.26 (12)	3.12±0.12 (12) $p=0.10$	4.86±0.24 (24) $p=0.05$	−0.75±0.19 (12) $p<0.001$	0.37±0.19 (12) $p<0.001$
			Hypophysectomized Rats		
Control	2.14±0.14 (14)	4.08±0.17 (8)	3.99±0.40 (16)	0.52±0.11 (16)	1.66±0.17 (10)
ACTH	1.92±0.11 (18)	3.97±0.12 (10)	4.39±0.20 (28) $p=0.3$	0.17±0.12 (18) $p=0.05$	1.39±0.21 (10)
ACE	1.23±0.13 (12) $p<0.001$	2.73±0.15 (12) $p<0.001$	4.68±0.25 (24) $p=0.2$	0.34±0.06 (12) $p=0.2$	1.15±0.08 (12) $p=0.02$

* Mean ± standard error. † Number of animals in parenthesis.
‡ Fisher's p values.

Similar experiments were carried out with *hypophysectomized* rats (Li and Kalman, 1949). Male rats 40 days old were hypophysectomized by the parapharyngeal approach and were employed 10 days later. All animals were allowed to eat the usual diet *ad libitum.* The hormone (5 mg) was administered intraperitoneally 24 hours before sacrifice. The glucose utilization and glycogen formation *in vitro* in the isolated diaphragm were carried out as previously described (Li *et al.*, 1949).

As shown in Table 3, the glucose uptake and glycogenesis in the muscle do not change after ACTH treatment. This is in contrast with the results obtained with the treated diaphragms of normal animals. In order to confirm this observation, hypophysectomized rats were treated with adrenal cortical extract (5 cc; Upjohn preparation) 24 hours before sacrifice. Results show that glycogenesis does not impair, but the extract definitely causes glycogenolysis (see Table 3).

It is therefore clear that the ACTH treatment has no influence on the glycogenesis of isolated diaphragm of *hypophysectomized* rats and that it inhibits the insulin effect in promoting glycogen storage in the diaphragms of *normal* rats. These observations suggest that ACTH induces the hypophysis of the normal animal to secrete another hormone (growth hormone?) which is responsible for the inhibiting effect.

ACTH Peptides

As mentioned earlier, certain peptic digests of the ACTH protein retain the adrenocorticotropic activity. In the following, the preparation and properties of the ACTH peptides are discussed.

Methods of Preparation. The adrenocorticotropic hormone was isolated from sheep glands by the procedure previously described (Li, 1947). The hormone appears to be a homogeneous protein as examined in electrophoresis, ultracentrifuge, and solubility studies. Approximately 10 mg of a commercial pepsin preparation (Lily) was dissolved together with 100 mg of the pure hormone in 20 cc 0.05*M* HCl solution; the mixture was kept at 37.4°C for 4 to 5 hours. Before 5 cc of 30% trichloroacetic acid was added to the digestion mixture, it was kept at 80°C for 3 minutes to destroy the peptic activity. In every case, it was found that the hormone was disgested to the extent of about 50% as determined by the nitrogen in the trichloroacetic acid supernatant. When the trichloroacetic acid supernatants were assayed in hypophysectomized female rats, results indicated that they contained the adrenocorticotropic activity (Li, 1948).

An alternative method for the preparation of the active peptic hydrolysate of ACTH employs the crystalline enzyme. One gram of ACTH protein and 40 mg crystalline pepsin were dissolved in 100 cc of 0.10*M* acetic acid and kept at 37.4°C. At the end of 17 hours, the enzyme action was stopped by putting the mixture into a 80°C water bath for 5 minutes. Seventeen cubic centimeters of 30% trichloroacetic acid was then added; the precipitate formed was removed by centrifugation. The supernatant was extracted with ether repeatedly until it was freed from thrichloroacetic acid. The non-protein fraction was then frozen and dried in vacuum. A typical assay data of this material as obtained by the ascorbic acid depletion method (Sayers *et al.*, 1949) may be seen in Table 4. It is clear that the trichloroacetic acid soluble fraction is highly active whereas the remaining protein residue contains practically no potency. This adrenocorticotropically active non-protein material has been employed for physicochemical and clinical investigations; it is hence designated as ACTH peptides.

TABLE 4

Bioassay of Fractions from Pepsin Hydrolysis of ACTH by Ascorbic Acid Depletion Method

Fraction	Dose in Nitrogen per 100 g rats (μg)	Ascorbic Acid Lowering per 100 g Adrenals (mg)	ACTH Standard Equivalent per μg N (μg)
L1642M-original ACTH protein	0.75 (4)*	116 (101-135)†	1.0
L1669CS-TCA‡ soluble material	0.597 (6)	121 (98-151)	1.5
L1669CP-TCA insoluble material	0.705 (5)	86 (62-138)	0.3

* Number of rats in parenthesis.
† Range of changes in parenthesis.
‡ TCA = trichloroacetic acid.

It should be mentioned that if the peptic hydrolysis was extended to 60%, a definite destruction of the hormonal activity was evident. Other enzymes have been investigated and it was found that the ACTH activity was destroyed by trypsin, chymotrypsin, and papain digestions. As shown earlier (Li, 1948), certain HCl hydrolysates of the hormone also retained adrenocorticotropic potency.

TABLE 5

Relative Peptide Size of the ACTH Peptides from Pepsin Digests

% of Digestion	Total N of Hydrolysate (mM)	NH_2-N (mM)	N in Peptides (mM)	Average Peptide Length
48	1.04	0.111	0.830	7.5
48	1.09	0.099	0.865	8.7
50	1.07	0.093	0.850	9.1
51	1.07	0.113	0.850	7.5

Relative Size of ACTH Peptides. The ACTH peptides do not contain free amino acids as indicated by the gasometric amino acid carboxyl determinations (Van Slyke *et al.*, 1941). It was found by the nitrous acid method of Van Slyke (1929) that approximately 10% of the total nitrogen in the peptides is free amino nitrogen. As previously reported (Li, 1949b), the complete hydrolysates of ACTH contained 6.5% of the total nitrogen as amide nitrogen and 79.4% of the total nitrogen minus the amide nitrogen as the amino nitrogen. With these data, it was computed that the average peptide length in the ACTH peptide varies from 7 to 9. The data are shown in Table 5.

Average Molecular Weight. In order to estimate the average molecular weight (M) of the peptides, we employed the well-known Svedberg formula

$$M = \frac{RTs}{(1 - V\rho)D} \qquad (1)$$

where s is the sedimentation constant, V the specific volume, D the diffusion constant, ρ the density of the solvent, R the gas constant, and T the absolute temperature. In collaboration with Kai O. Pedersen at Upsala (1950) we have obtained an average value for the sedimentation constant, s_{20}, to be

TABLE 6

Estimated Average Molecular Weights for ACTH Peptides for Different Values of V_{20}

$s_{20} \times 10^{13}$	$D_{20} \times 10^{7}$	V_{20}	M	f/f_0
0.45	30.0	0.75	1450	(0.94)
		0.72	1300	(0.99)
		0.70	1210	1.02
		0.68	1140	1.05

0.45S and the diffusion constant, D_{20}, 30.0 $\times$ 10^{-7} (cgs units). In the case of the specific volume, we have not yet determined it on account of the limited supply of the peptides, but it is reasonable to assume a value for V somewhat in the neighborhood of 0.70. In Table 6, we have tabulated the calculated M from equation 1 using different values of V_{20}. Since the frictional constant, f/f_0, is always greater than unity, we assume that V must be less than 0.72. The average molecular weight of the ACTH peptides is therefore less than or equal to 1200.

TABLE 7

Certain Physicochemical Data of ACTH Peptides

Molecular weight	1200
Diffusion constant, $D_{20} \times 10^{7}$	30.0
Sedimentation constant, S_{20}	0.45
Average amino acid residues	8
N, %	13.2
NH_2-N, %	1.4

Certain Chemical Properties. A complete hydrolysate of the ACTH peptides has revealed the following amino acids on two-dimensional paper partition chromatography (Consden *et al.*, 1944) using lutidine/H_2O and phenol/H_2O as the solvents: aspartic acid, glutamic acid, lysine, arginine, serine, glycine, threonine, alanine, tyrosine, histidine, valine, tryptophane, proline, leucines, and phenylalanine. Quantitative analysis of tyrosine and tryptophane content by the method of Lugg (1937) showed that the peptides contain 1.0% tryptophane and 1.5% tyrosine.

The peptides are very soluble in water. The biological activity of a neutral solution is stable after being kept at 100°C for 2 hours. However, it is destroyed by further hydrolysis with 6 N HCl at 37.4°C.

Table 7 summarizes physicochemical properties of the ACTH peptides.

Preliminary Chromatographic Studies. The peptides may be separated into a number of components on paper partition chromatography using either 10% acetic acid-butanol/H_2O or phenol/H_2O as the solvent. Whatman no. 4 paper was used. Results indicated that the peptides contain at least six distinct spots using 10% Hac-BuOH/H_2O after being developed by ninhydrin solution. It was found that one of these spots was highly active in adrenal-stimulating potency as assayed by the ascorbic acid depletion method in hypophysectomized rats. Similar experiments using phenol/H_2O resulted only in 3 ninhydrin spots. The spots were cut out and eluted with water; the resultant solutions were assayed for the adrenocorticotropic activity. As shown in Table 8, one of the spots is highly active. It is evident from such limited assay data that the active spot is approximately two times more active than the original peptide mixture.

TABLE 8

Bioassay of Peptide Fractions Obtained in Paper* by Ascorbic Acid Depletion Method

Fraction	N Distribution (%)	Dose in Nitrogen per 100 g rats (μg)	Ascorbic Acid Lowering per 100 g Adrenal (mg)	ACTH Standard Equivalent per μg N
L1563 D2—original peptides	100	0.75(4)†	125 (−105−139)‡	1.3
"Active Spot"	18	0.81(6)	142 (103−176)	2.1
Remainder of the paper	72	3.24(8)	3 (+21−51)	0

* Using phenol/H_2O as the solvent.
† Number of rats in parenthesis.
‡ Range of changes in parenthesis.

The starch column chromatography of Stein and Moore (1948) was also employed for the fractionation of the ACTH peptides. The eluent was a mixture consisting of 1 part of 0.1 *N* HCl, 2 parts of *n*-propanol, and 1 part of *n*-butanol. When the fractions were assayed, the first fraction collected from tubes* No. 15 to 20 was found to contain most of the adrenocorticotropic activity. We are not certain, however, of the purity of this fraction. The fact that similar peptides were obtained by Borsook *et al.,* (1949) from livers of different animals suggests that our adrenocorticotropically active peptide fraction obtained from the starch column must contain a mixture of peptides.

Clinical Studies. In collaboration with Rolf Luft at Stockholm, we have investigated the metabolic effect† of the ACTH peptides in a patient (a male, 58 years old) suffering from severe rheumatoid arthritis using a daily dose

* Each tube contained 0.5 cc effluent.

† Earlier studies in collaboration with L. W. Kinsell in California have already indicated that 100 mg daily dose of ACTH peptides was effective to such treatment.

of only 13 mg (divided into six injections). The improvement was evident within 3 to 4 hours after an injection of 2 mg of the peptides. A similar benefical effect was also observed when the ACTH protein was used (Luft *et al.,* 1949).

There were no differences between ACTH peptides and ACTH protein in their ability to stimulate the adrenal cortex as shown by the retention of sodium, chlorides, water, by the increase of the excretion of 17 ketosteroids and uric acid, and by the disappearance of circulating eosinophil leucocytes.

Summary

Methods for the preparation of a peptide mixture from the peptic digest of ACTH protein are described. It is demonstrated that the ACTH peptides contain an average of eight amino acids and have an average molecular weight of 1200. It is highly active in the treatment of rheumatoid arthritis. The peptides may be fractionated by chromatographic techniques. When the ACTH peptide is isolated in pure form and its structure is elucidated, the total synthesis of this important hormone might well be practicable.

The new knowledge of the nature of ACTH protein is also discussed. It is shown that it has only one polypeptide chain with alanine as the end group. The fact that the free amino group is essential for the adrenocorticotropic activity is confirmed. Studies of glycogenesis in the isolated diaphragm show that ACTH treatment has no influence in hypophysectomized rats, but it inhibits the insulin effect in promoting glycogen storage in the muscle of a normal animal.

REFERENCES

Borsook, H., Deasy, C. L., Haagen-Smith, A. J., Keighley, G., and Lowy, P. H., *J. Biol. Chem.,* **119,** 705 (1949).

Consden, R., Gordon, A. H., and Martin, A. G. P., *Biochem. J.,* **38,** 224 (1944).

Herriott, R. M., *J. Gen. Physiol.,* **19,** 283 (1935).

Li, C. H., *Ann. Rev. Biochem.,* **16,** 291, (1947).

Li, C. H., *Transactions of Macy Conference on Metabolic Aspects of Convalesence* (Josiah Macy Jr. Foundation), **17,** 114 (1948).

Li, C. H., Abstractions of Communications of First International Congress of Biochemistry, Pp. 386 (1949a).

Li, C. H., *Federation Proc.,* **8** (March, 1949b).

Li, C. H., Simpson, M. E., and Evans, H. M., *Arch. Biochem.,* **9,** 259 (1946).

Li, C. H., and Evans, H. M., *Vitamins and Hormones,* **5,** 197 (1947).

Li, C. H., and Kalman, C., unpublished data, 1949.

Li, C. H., Kalman, C., and Evans, H. M., *Arch. Biochem.,* **22,** 357 (1949).

Li, C. H., and Porter, R. R., unpublished data, 1949.

Li, C. H., and Pedersen, Kai O., *Arkiv Kemi,* **1,** 533 (1950).

Luft, Rolf, Sjögren, B., and Li, C. H., *Acta Endocrinologica,* **3,** 299 (1949).

Lugg, J. W. H., *Biochem. J.,* **32,** 775 (1937).

Sanger, F., *Biochem. J.,* **39,** 507 (1945).

Sayers, M. A., Sayers, G., and Woodbury, L. A., *Endocrinology,* **42,** 379 (1949).

Stein, W. H., and Moore, S., *J. Biol. Chem.,* **176,** 337 (1948).

Van Slyke, D. D., *J. Biol. Chem.,* **83,** 425 (1929).

Van Slyke, D. D., Dillon, R. T., MacFadyen, D. A., and Hamilton, P., *J. Biol. Chem.,* **141,** 629 (1941).

SERUM CHOLESTEROL: A PROBABLE PRECURSOR OF ADRENAL CORTICAL HORMONES*

JEROME W. CONN, WILLIAM C. VOGEL, LAWRENCE H. LOUIS, AND STEFAN S. FAJANS,

WITH THE TECHNICAL ASSISTANCE OF

JANE BLOOD, BETTY SPRUNGER, AND BETTY JOHNSON

University of Michigan Medical School, Ann Arbor

The high concentration of cholesterol in the adrenal cortex has been well established. The sharp fall in adrenal cholesterol under conditions of increased cortical activity has been demonstrated by Long and his group (Long, 1947). Conversion of cholesterol to cortical hormones is believed to explain the phenomenon. In their acute experiments in rats, the Yale group (Long, 1947; Sayers *et al.,* 1945) reported no significant alterations in the cholesterol levels of other tissues (brain, liver, kidney, spleen, heart, skeletal muscle, lymphoid tissue, and plasma). As one reviews their data, however, he finds a rather consistent decrease of plasma cholesterol of about 20%.

The capacity of the adrenal cortex, under stimulation, to deliver large amounts of steroidal hormones into the blood seems clear (Vogt, 1943, 1943; Ingle, 1944). Under conditions of *intense* and *continued* stimulation, the source of large amounts of cholesterol to satisfy continuing adrenal requirements became of considerable interest. It is to be recalled that *in vitro* synthesis of cholesterol from acetate by surviving adrenal cortical tissue has recently been demonstrated by Chaikoff and his associates (Srere *et al.,* 1948).

The data given below appear to indicate that the *blood* constitutes an important source of cholesterol to the adrenal cortex when the gland is forced to produce large amounts of steroidal hormones for long periods of time. Of special interest is the fact that the esterified fraction, as compared with the free cholesterol of the serum, falls selectively and severely. This suggests that the ester cholesterol rather than the free fraction is the immediate precursor of the cortical hormones, especially since 90% of the cholesterol normally found in the cortex is in the ester form. No other tissue of the body is known to approach this ratio of ester to free cholesterol except (1) the blood and (2) the corpus luteum, another producer of steroidal hormones.

Methods

All the subjects were maintained on constant diets for metabolic balances throughout the studies. The data were obtained in the course of 9 separate experiments performed on 3 normal men, 2 male cases of Addison's disease,

* This work represents part of a project supported by the Research Grants and Fellowships Division of the U. S. Public Health Service. The paper was presented in part at the Atlantic City meeting of the Association for the Study of Internal Secretions, June, 1949. Published in full *J. Lab. Clin. Med.,* **35,** 504-517 (1950).

TABLE 1

Subject	Age	Sex	Weight (kg)	Diet Pro. (g)	Diet CHO (g)	Diet Fat (g)	Diet Cals	Control Period (days)	ACTH Period (days)	Total ACTH* (Armour std mg)	Recovery Period (days)	Remarks
R. S. Normal	37	M	73.8	98	300	197	3365	4	6	408	8	
R. S. Normal	37	M	69.7	98	300	197	3365	7	8	416	6	
R. S. Normal	37	M	70.5	98	300	197	3365	5	10†	2,000†	15	For comparison with ACTH in same subject†
G. A. Normal	22	M	74.4	108	300	173	3170	6	6	600	5	
D. E. Normal	24	M	64.7	98	300	176	3177	4	8	234	6	
R. D. Cushing's	20	F	62.7	95	252	92	2216	5	9	900	10	Bil. cortical hyperplasia at operation
V. K. "Cured" Cushing's	20	F	63.4	79	299	96	2376	5	5	500	5	Subtotal adrenalectomy (4 mos. post-op.)
F. H. Addison's disease	43	M	65.9	103	289	158	2990	6	5	510	0	No DCA during entire experiment
L. W. Addison's disease and diabetes	41	M	57.0	99	249	124	2508	6	6	550	0	DCA-3 mg per day and no insulin during entire experiment

* We are indebted to Dr. John R. Mote and Armour and Company, Chicago, for the purified ACTH used in these studies.
† Cortisone 200 mg/day. No ACTH.

1 female apparently cured of Cushing's syndrome by a 90+% subtotal adrenalectomy 4 months previously, and another female with active Cushing's syndrome. Data concerning the subjects and the procedure of each of the 9 studies are tabulated in Table 1.

Duplicate determinations of total and free cholesterol of serum were performed by the Schoenheimer-Sperry method as modified by Sperry (1945). Urinary 17-ketosteroids (Robbie and Gibson, 1943) and "11-oxysteroids" (Daughaday, 1948) were done daily.

Results

Figure 1 shows an example of the response in normal men. Blood was drawn three times daily throughout most of the experiment for determination of serum cholesterol.

The data are charted to show the actual fall of the total and free cholesterol of serum from their average baseline values. The black area between these two values represents the actual decrease of the ester fraction. For better orientation to the system used to depict these changes, the absolute values for this first experiment are shown in Fig. 2.

It is quite evident at a glance (Figs. 1 and 2) that between the third and fifth days of administration of ACTH a great decrease of total serum cholesterol has occurred. At the point of maximal decrease this amounts to a fall of 91 mg per 100 cc of serum (39.4% of 231 mg). It is clear, too, that of the 91 mg per 100 cc of serum which disappeared 79 mg was ester cholesterol (46.2% of 171 mg) and only 12 mg was free cholesterol. It is to be noted next that free cholesterol maintained its baseline value during the first three days of administration of ACTH, but that the ester fraction had begun to fall on the morning of the second day of ACTH. When the precipitous decrease of the ester fraction occurred, it was followed by a decrease of the free fraction, but the latter fall was of much smaller proportions, both in absolute and in relative terms. Upon cessation of ACTH free cholesterol bounded above its baseline value in one day. On the other hand, it required six days for the ester fraction to return to its normal value.

The data on urinary excretion of 17-ketosteroids and of "11-oxysteroids" are included merely to indicate that the quality and quantity of ACTH employed in these experiments were such as to produce intense and prolonged stimulation of adrenal corticosteroid production.

Figure 3 represents the results of a similar experiment done upon a patient with active Cushing's syndrome. It is seen that the decrease of serum cholesterol is qualitatively the same as that observed in the normals. Quantitatively, however, the response is much greater. The maximal fall of the ester fraction was 115 mg per 100 cc and that of the free cholesterol was 45 mg per 100 cc. Note again the delay of several days before the ester fraction falls sharply.

Figure 4 represents a similar experiment performed upon a patient with Addison's disease. His daily dose of DCA had been discontinued at the beginning of his baseline period, seven days prior to the first day of ACTH. It is noteworthy, first, that no change in the excretion of 17-ketosteroids oc-

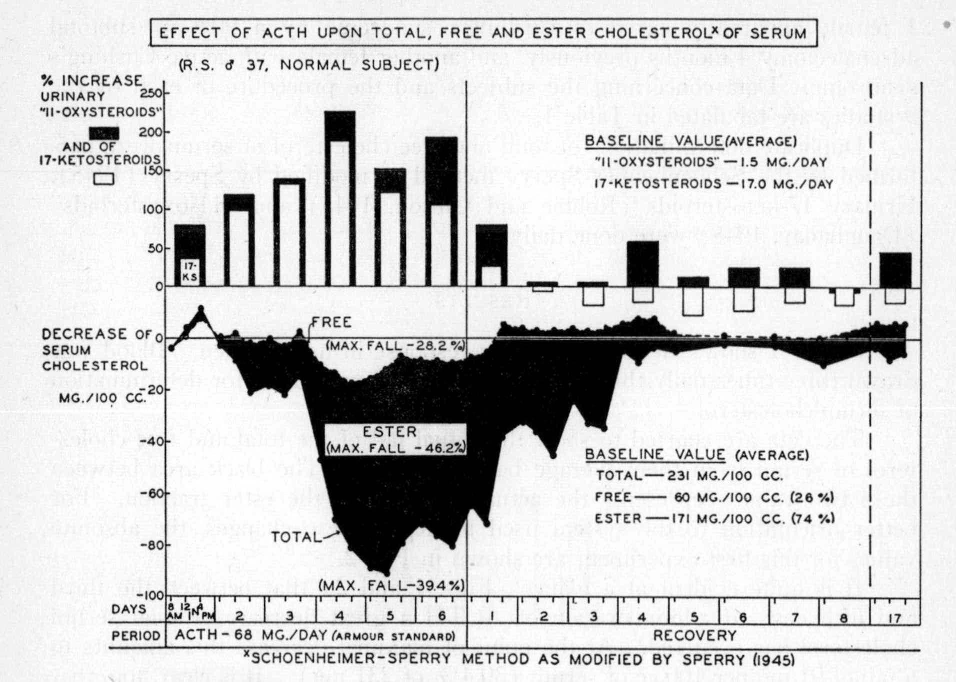

FIG. 1. Effect of ACTH upon serum cholesterol (normal subject).

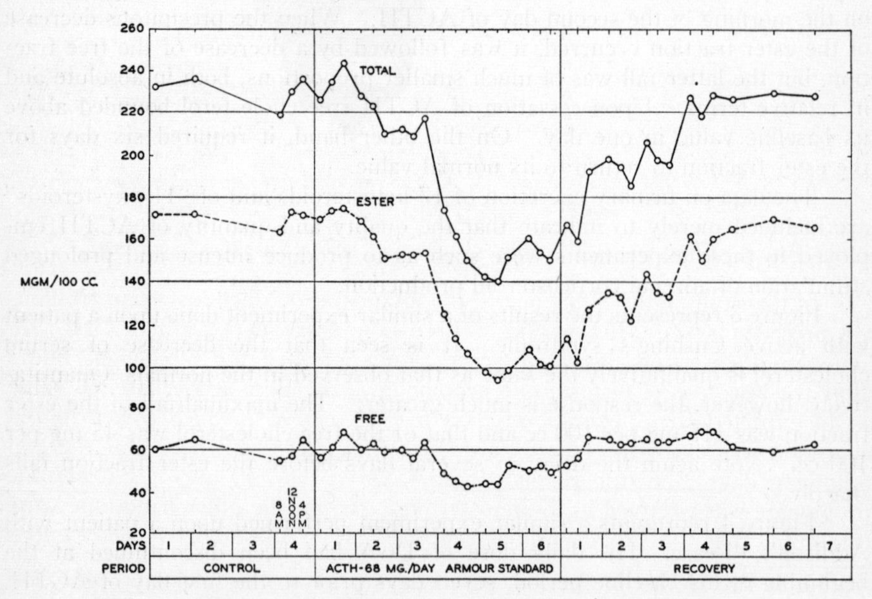

FIG. 2. Effect of ACTH upon serum cholesterol (normal subject).

curred during a five-day period of large doses of ACTH; secondly, that during this period "11-oxysteroid" excretion fluctuated between 50% above and 50% below the low baseline values*; and, thirdly, that no significant changes of serum cholesterol occurred.

Figure 5 reveals an interesting set of data. These were obtained upon another case of Addison's disease who in addition had diabetes mellitus. In this study DCA (3 mg per day 1 M) was administered throughout the entire experiment. Insulin was discontinued at the beginning of the baseline period. It is observed that the baseline value of 17-ketosteroid excretion is at the expected value for male Addison's disease. However, baseline "11-oxysteroid" excretion is well within the normal range. A possible *modus operandi* for such dissociations of adrenal cortical function in Addison's disease will be discussed elsewhere.

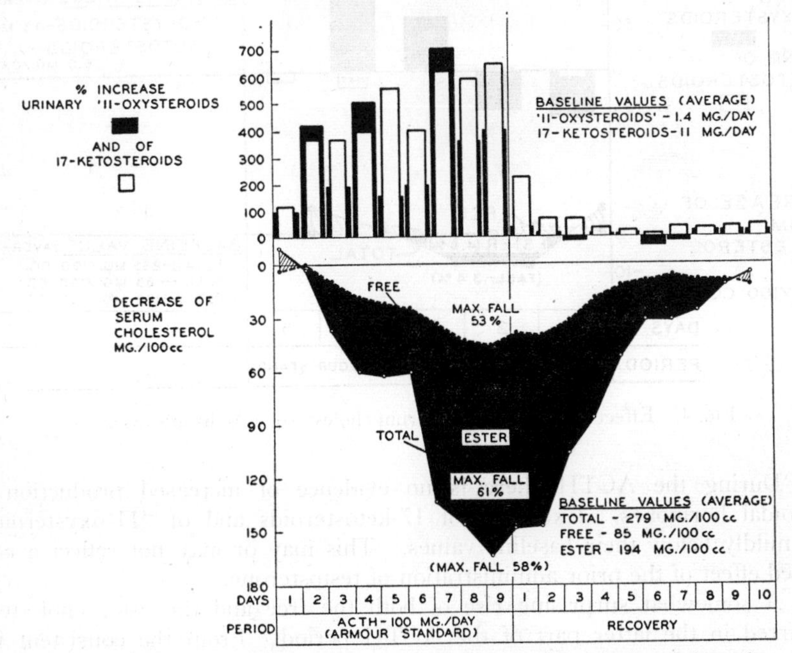

FIG. 3. Effect of ACTH upon serum cholesterol (Cushing's syndrome).

In this study a three-day period of administration of testosterone propionate preceded the ACTH period. Three points are noteworthy. First, 17-ketosteroid excretion increased greatly and persisted for forty-eight hours after the last dose of testosterone propionate. If one uses the baseline value as the expected rate of excretion for the five-day period in which 17-ketosteroid excretion was excessive (three days of testosterone propionate plus the two following days), he can calculate that 43% of the 150 mg of testosterone propionate was excreted as 17-ketosteroids. If, on the other hand, one were to assume that the exogenous testosterone depressed completely endogenous

* It is also worthy of mention that circulating blood eosinophil cells done daily showed no significant deviation from the baseline values.

production of 17-ketosteroids precursors and that all the excreted 17-ketosteroids were derived from the testosterone propionate administered, calculation would indicate a 55% conversion of this compound to excretory 17-ketosteroids. That the value will fall somewhere between 43% and 55% is obvious. This is in close agreement with the figure of 50% observed by others and indicates the normal capacity of this subject to carry out this conversion. The second point of interest is the apparent reduction of excretory "11-oxysteroids" in the testosterone period. Thirdly, no significant change in serum cholesterol was observed during the administration of testosterone propionate.

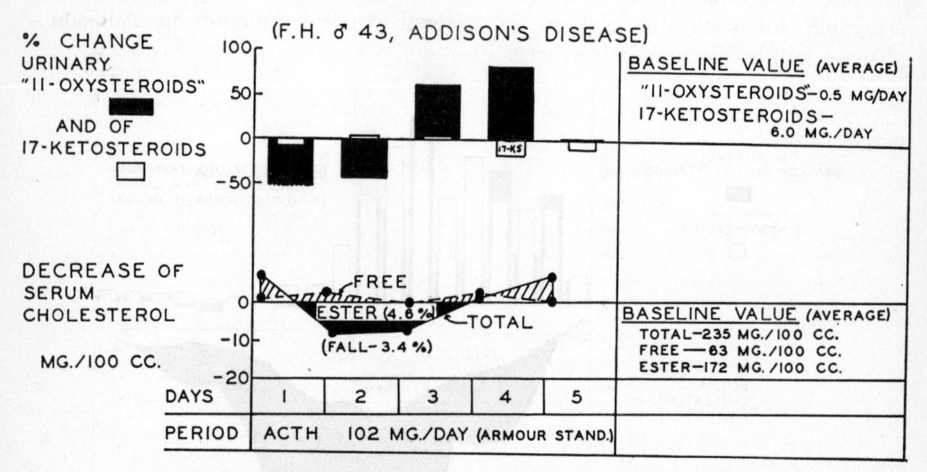

FIG. 4. Effect of ACTH upon serum cholesterol (Addison's disease).

During the ACTH there is no evidence of increased production of steroidal hormones. Excretion of 17-ketosteroids and of "11-oxysteroids" are mildly below their baseline values. This may or may not reflect a continued effect of the prior administration of testosterone.

A somewhat surprising rise of both the free and the ester cholesterol occurred in the latter part of the ACTH period. From the consistent fall of serum cholesterol exhibited by normal persons given ACTH and from the lack of any change observed in the prior case of Addison's disease given ACTH, it seems likely that the rise of serum cholesterol observed in this patient is related to the presence of diabetes. Insulin had been stopped eleven days before ACTH was begun. If the administered ACTH was responsible for this rise of serum cholesterol the mechanism is obscure at present. In any case, serum cholesterol did not fall as it does in the presence of functioning adrenals.

Figure 6 shows the results obtained when 200 mg per day of cortisone (in place of ACTH) was administered to the same normal man who had been studied previously with ACTH (Figs. 1 and 2). There appears to be a mild but not well-maintained depression of serum cholesterol. Whether or not it is real we are not yet prepared to say. During the entire experiment only

3 points fell below the lower limit of the range of baseline values. It may be significant that the most intense *metabolic effects* of cortisone were observed during the five-day period immediately following *cessation* of cortisone. Yet during this period serum cholesterol, if it was changing significantly, was rising. On the other hand, the higher than baseline values of the latter part of the recovery period suggest that the mild depression observed earlier *might* have been real.*

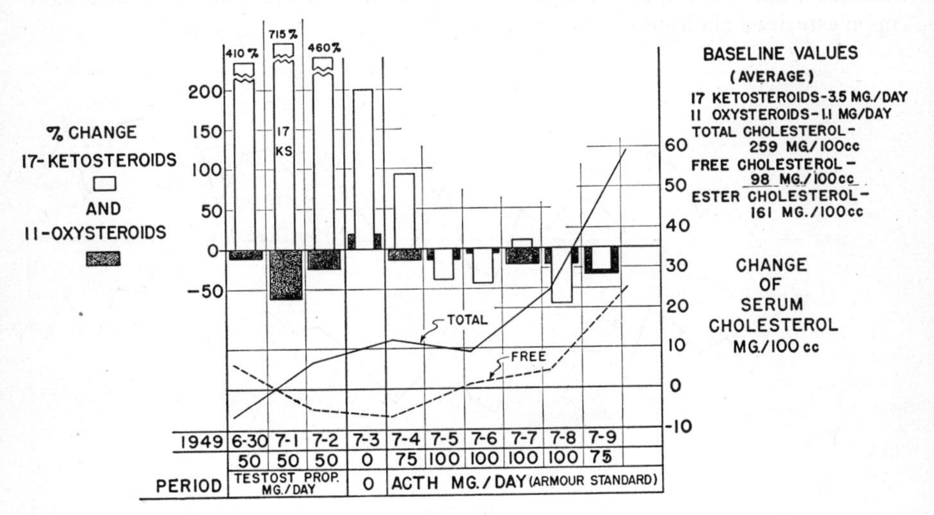

FIG. 5. Effect of ACTH upon serum cholesterol (Addison's disease with coexisting diabetes mellitus).

DISCUSSION

The data demonstrate that persistent stimulation of the adrenal cortices of man by administration of ACTH results in a very significant decrease of total circulating serum cholesterol, and that upon cessation of ACTH, serum cholesterol returns to its baseline concentration in four to five days. On the other hand, similar treatment with ACTH of patients with Addison's disease produces no lowering of serum cholesterol. One can conclude that the adrenal cortices must be capable of greatly accelerated functional activity if the decrease of serum cholesterol is to occur. The possibility that the phenomenon might be produced by a thyrotropic contaminant in the ACTH is eliminated by the absence of any fall in serum cholesterol in the ACTH-treated Addisonians.

The fact that the fall of serum cholesterol is dependent upon an intense and continued response of the cortex to ACTH limits speculation regarding its mechanism to two major possibilities: (1) that the decrease of serum

* Since completion of this report we have followed total and free cholesterol on another subject (rheumatoid arthritis) who received 200 mg per day of cortisone for ten days. No decrease of cholesterol was observed. In fact, serum cholesterol rose mildly during the period of cortisone activity.

cholesterol is due to a specific activity of one or more of the cortical steroids which are produced in excess or (2) that under conditions of forced production of large quantities of steroidal hormones, the adrenal cortices, after having depleted their own cholesterol reserves, use the blood as a further source of the compound for continued synthesis of steroidal hormones. The differences observed between the cholesterol ester fraction and the free fraction with respect to time that each begins to decrease, their rates of descent, their maximal falls, and their rates of return to the baseline, focus particular attention upon esterified cholesterol, the prime mover.

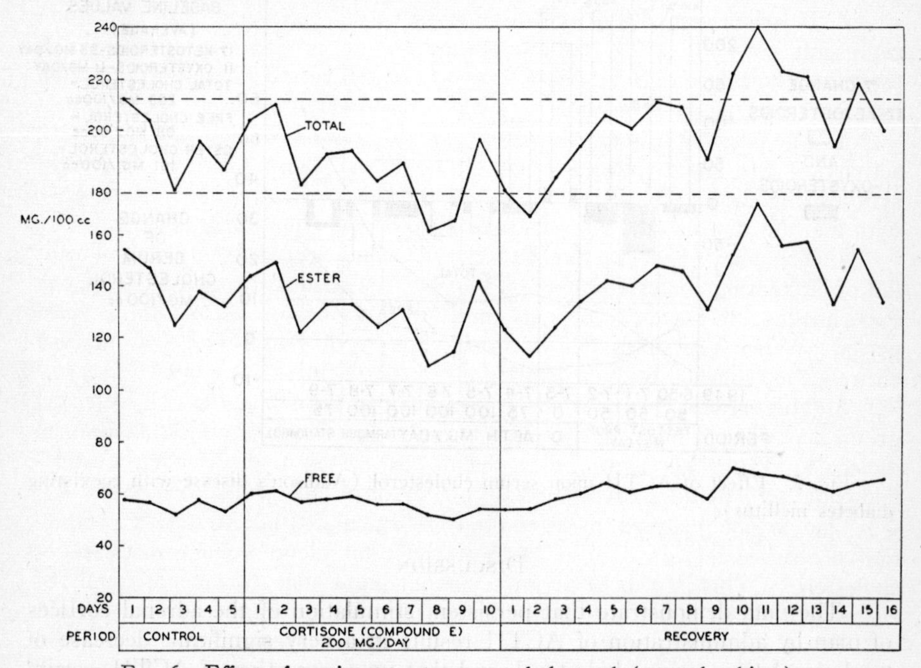

FIG. 6. Effect of cortisone upon serum cholesterol (normal subject).

Insufficient data are available to rule out with certainty a pharmocological effect of adrenal steroids. In a few isolated experiments blood cholesterol has been followed during administration of compound E. Sprague during studies in Addison's disease gave the material on two occasions in doses of 20 mg per day and 100 mg per day respectively. No decrease of serum cholesterol was observed. More recently he found no significant changes of serum cholesterol in patients with rheumatoid arthritis given 200 mg per day of cortisone.* Perera *et al.*, (1949) noted "a slight decrease in cholesterol esters" in three hypertensives given cortisone (80 mg per day).

It may be that 400 or 500 mg per day of cortisone will produce a change in serum cholesterol similar to that which we have observed with 50 mg per day of ACTH. If the change *were* due to a peripheral effect of the steroids it would imply one of two mechanisms, either decreased hepatic synthesis of

* See preceding footnote.

cholesterol, particularly of ester cholesterol or increased removal of cholesterol from the blood for the purpose of oxidation, fat transport, or tissue desposition. Little is known of these possibilities in relation to increased cortical function.

Favoring removal of ester cholesterol from the blood for the purpose of conversion to steroidal hormones during intense and prolonged adrenal hyperactivity, are the following points:

1. ACTH, epinephrine or stress produce a sharp fall in total adrenal cholesterol (Long, 1947). Over 90% of that which disappears from the gland is ester cholesterol.

2. If it is accepted that the disappearance of cholesterol from the *gland* represents conversion to steroidal hormones, the predominant precursor is esterified cholesterol. That cholesterol *can* be converted to another steroidal hormone, pregnanediol has already been demonstrated by Block (1945).

3. The fact that adrenal cholesterol *falls* sharply with increased functional activity of the gland indicates that although the gland may be capable of synthesizing cholesterol (Srere *et al.*, 1948), this process is not keeping pace with the rate at which the material is being removed from the gland.

4. The delay of several days, which we have observed, before a significant fall of circulating cholesterol occurs, suggests that under the conditions of our experiments, the cortex is able to supply the needed precursors for a brief period. When the sharp fall of blood cholesterol *does occur,* it is almost entirely the esterified fraction which falls.

5. The organic chemist finds it necessary to block the reactive hydroxyl group on carbon-3 when he attempts to make substitutions on the cholesterol molecule. From this point of view the adrenal cortex might well select esterified cholesterol as the more immediate precursor of its steroidal hormones.

One is not justified in attempting to calculate the presumed production of corticosteroids on the basis of the total blood volume decrease of cholesterol. This is so because one does not know that the hepatic supply of cholesterol to the blood is constant under these circumstances. There might well be an increasing supply to the blood as the concentration in the blood diminishes. If, however, one assumes a constant rate of supply of cholesterol to the blood under the conditions of these experiments, rough calculation gives the following interesting figures. They are listed in the order of increasing intensity of overall metabolic responses to ACTH.

Subject	Overall Metabolic Response (+ to ++++)	Dose of ACTH mg/day	Calculated Conversion of Cholesterol to Corticosteroids mg/day
F.H. (Addison's)	0	100	0
L.W. (Addison's)	0	100	0
V.K. (Subtotal Ad x)	+	100	280
D.E. (Normal)	++	39	275
G.A. (Normal)	+++	100	426
R.S. (Normal)	++++	50	650
R.S. (Normal)	++++	68	1138
R.D. (Cushing's)	++++	100	1031

These are very crude calculations, probably representing only distant approximations of daily corticosteroid production. Nevertheless, the figures produced are of an order of magnitude that seems reasonable.

Finally, speculations such as those embarked upon above necessarily lead to an appraisal of hepatic function as a possible limiting factor in the production of adrenocortical steroids. If the fall of esterified cholesterol of serum represents withdrawal from the blood of this substance for synthesis of adrenal steroids, the source of supply would be dependent predominantly upon the liver. A decreased capacity to esterify cholesterol in normal amounts, as occurs in severe hepatic insufficiency, might account for diminished adrenocortical responses to stress, especially if the stress were prolonged. This critical problem is now being studied.

Summary and Conclusions

1. Normal people given large amounts of ACTH exhibit a sharp decrease of total serum cholesterol after several days of such treatment. The baseline level is regained three to five days after cessation of ACTH. A similar but more intense effect was observed in ACTH-treated Cushing's syndrome.

2. The esterified fraction of serum cholesterol falls sooner than the free fraction, and its percentage decrease from its baseline value is significantly greater than that which occurs in the free fraction. Eighty-five per cent of the decrease of total cholesterol is due to disappearance of the ester fraction.

3. Upon cessation of ACTH free cholesterol returns to its baseline value before the esterified fraction regains its baseline value. At this point in the recovery phase the deficiency of total serum cholesterol is due wholly to a diminished amount of esterified cholesterol.

4. Patients with Addison's disease treated with ACTH in a similar way show no significant decreases in either free or esterified serum cholesterol.

5. The data may be interpreted in one of two ways: first, that the fall of serum cholesterol is the result of the physiological activities of one or more of the cortical steroids produced in excess when ACTH is administered or, secondly, that the blood constitutes a source of cholesterol for the synthesis of cortical hormones after the cortex has depleted its cholesterol reserves in the course of prolonged ACTH stimulation, the latter forcing continued production of large amounts of steroidal hormones.

6. In the light of present knowledge and until more data are available on the specific effects of cortical steroids upon serum cholesterol, the second possibility is favored as best explaining the data.

7. Based upon the assumption that esterified cholesterol of serum is drawn upon during periods of acute or chronic stress as a readily available precursor with which to satisfy the increased demand for corticosteroid production, diminished hepatic function may assume the role of a limiting factor in the total response of the cortex to stimulation.

REFERENCES

BLOCH, K., The biological conversion of cholesterol to pregnanediol. *J. Biol. Chem.,* **157,** 661 (1945).

DAUGHADAY, W. H., JAFFE, H., and WILLIAMS, R. H., Clinical assay for "Cortin." *J. Clin. Endocrinol.,* **8,** 166 (1948).

INGLE, D. J., "The Physiological Action of Adrenal Hormones," in *Chemistry and Physiology of Hormones,* pp. 83-103. American Association for the Advancement of Science, Washington, 1944.

LONG, C. N. H., "The Relation of Cholesterol and Ascorbic Acid to the Secretion of the Adrenal Cortex," in *Recent Progress in Hormone Research,* Vol. I, pp. 99-122. Academic Press, New York, 1947.

PERERA, G. A., PINES, K. L., HAMILTON, H. B., and VISLOCKY, K., Clinical study of 11-dehydro-17-hydroxy-corticosterone in hypertension, Addison's disease and diabetes. *J. Clin. Invest.,* **28,** 803 (1949).

ROBBIE, W. A., and GIBSON, R. B., Rapid clinical determination of 17-ketosteroids. *J. Clin. Endocrinol.,* **3,** 200 (1943).

SAYERS, G., SAYERS, M. A., LIANG, T. Y., and LONG, C. N. H., The cholesterol and ascorbic acid content of the adrenal, liver, brain and plasma following hemorrhage. *Endocrinology,* **37,** 96 (1945).

SPERRY, W. M., The Schoenheimer-Sperry Method for Determination of Cholesterol. Department of Biochemistry, New York Psychiatric Institute and Hospital, New York, Revised April, 1945.

SPRAGUE, R. G., personal communication.

SRERE, P. A., CHAIKOFF, I. L., and DAUBEN, W. G., The *in vitro* synthesis of cholesterol from acetate by surviving adrenal cortical tissue. *J. Biol. Chem.,* **176,** 829 (1948).

VOGT, M., The output of cortical hormone by the mammalian suprarenal. *J. Physiol.,* **102,** 341 (1943).

VOGT, M., Observations on some conditions affecting rate of hormone output by the suprarenal cortex. *J. Physiol.,* **103,** 317 (1944).

FACTORS REGULATING THE ADRENAL CORTICAL SECRETION

C. N. H. LONG

Yale University, New Haven, Connecticut

In considering the manner of regulation of the secretion of the adrenal cortex there is one basic fact that cannot be emphasized too often. It is that an increased rate of secretion of at least the 11-oxy adrenal cortical steroids is only possible following a preliminary release of adrenocorticotropic hormone (ACTH) from the anterior pituitary. In other words, any study of the regulation of the adrenal cortical secretion to a large degree resolves itself into a study of the factors responsible for the release of ACTH from the anterior pituitary. It is presumed that to this knowledge we will one day have to add more information concerning the factors, if any, that regulate the formation of ACTH in the gland since those that govern its release are not necessarily synonymous with those that govern the formation of the hormone in the gland.

The importance of acquiring more basic knowledge of the mechanisms governing the secretion of ACTH and the adrenal cortical hormones (ACH) has become increasingly evident with the publication of studies that show that either ACTH or ACH are highly effective in alleviating a number of clinical disorders, particularly rheumatic fever, rheumatoid arthritis, and the allergic states. However, since both these agents, cortisone and ACTH, are likely to be exteremely scarce for an indefinite period, it is necessary at this time to consider the use of other agents which may be capable of affording at least partial relief in these situations by provoking the secretion of ACTH from the individual's own pituitary gland.

Methods for the Detection of an Increased Rate of Adrenal Cortical Secretion

A variety of methods have been used to detect an increased secretory rate of the adrenal cortical steroids. These include (1) the direct or indirect measurement by chemical or biological methods of the content of those hormones in the blood in the adrenal vein, (2) the detection of cortical hormones in lipid extracts of the urine either by chemical or biological techniques, (3) the fall in adrenal ascorbic acid or cholesterol which is a specific response of the gland to an increased blood level of ACTH, (4) the fall in the number of circulating eosinophils or lymphocytes which appears to be a specific response to an increased blood level of adrenal cortical hormones.

It is obvious that the choice of method will be determined by the type of experimental animal and by the rapidity of measurement that is desired. In man examination of the urine or the change in circulating eosinophils or lymphocytes is the method of choice, whereas in experimental animals these same changes in the white blood cells or the fall in adrenal ascorbic acid or cholesterol are the most reliable and simplest techniques. In our experiments in which rats were the experimental animals we have used these two methods and have been able to show that there is a high degree of correlation between the fall in circulating eosinophils and the decline in adrenal ascorbic acid. That such a correlation exists is of some importance since the fall in circulating eosinophils is a measure of the blood level of adrenal steroids, while the decline in adrenal ascorbic acid is a measure of the blood level of ACTH. That such a correlation should exist is in keeping with the known relationship between these two glands of internal secretion.

Conditions Associated with an Increased Rate of Adrenal Cortical Secretion

A large number of circumstances are known, both in man and in experimental animals, to cause an activation of the anterior pituitary-adrenal cortical mechanism as measured by the methods outlined above.

Since an increased secretion of adrenal cortical hormones is conditional upon a prior secretion of ACTH, two main hypotheses have been suggested to account for this activation of the anterior pituitary by such a wide variety of initiating causes.

Humoral (Sayers and Sayers, 1948). This envisages a reciprocal relationship between the blood levels of ACH and ACTH of a kind so that a decline in blood ACH serves as the stimulus for ACTH secretion, and conversely a rise in blood ACH inhibits the secretion of ACTH. In order that this theory may have any weight it is necessary to suppose that *all* conditions associated with increased secretion of ACH are first preceded by an *increased "utilization"* of ACH by the cells of the body, which is followed by a fall in the blood level of ACH. This in turn stimulates the release of additional ACTH tending to restore the blood level of ACH to normal.

This theory is supported by the well-substantiated fact, first demonstrated by Sayers, that the prior injection of cortical hormones inhibits the response of the anterior pituitary to circumstances which normally invoke the release of ACTH. In practice it will be found that this inhibitory effect of exogenous ACH is not an absolute but a quantitative one since very strong stimulation will still cause release of ACTH as measured by the level of adrenal ascorbic acid.

The question as to whether there is an actual "utilization" of cortical hormone by the tissues is still unproved. So far it has not been shown (1) that there is an arterio-venous difference in blood level and (2) that this A-V difference, if it exists, is increased by those circumstances that are known to augment the rate of secretion of adrenal cortical hormones.

Furthermore, it must be stated that the rapid decline in circulating eosinophils or adrenal ascorbic acid that follows brief stimulation of sensory nerves, or the mere fright that often follows attempts to handle timid animals is difficult to explain merely on the grounds that the tissue "utilization" of cortical hormone has been increased so rapidly or to such a degree by such trivial procedures.

Nevertheless, I believe it should be stated at this time that provided it can be demonstrated that the tissue "utilization" of cortical hormone is increased by procedures known to activate the adrenal cortex—and this should be possible, then this humoral hypothesis provides an explanation for the *long-continued* stimulation of the secretion under a variety of circumstances. For it is apparent that there are many situations where the demands for extra cortical hormone continue for long periods and under these conditions a humoral mechanism to ensure the continued secretion would appear to be a logical one.

Neuro-humoral. The observation that brief stimulation of sensory nerves, such as is produced by cutting the skin or the subcutaneous injection of a few drops of an irritating fluid such as 10% NaCl solution, is followed by a rapid decline in adrenal ascorbic acid or circulating eosinophils raises the question of a neural or neuro-humoral factor in the regulation of the ACTH secretion of the anterior pituitary.

A survey of the circumstances under which the rapid release of ACTH occurs indicates that the majority of them have previously been shown to be those that are also associated with the discharge of epinephrine from the adrenal medulla. These circumstances are too numerous to mention, but range all the way from changes in the physical environment to the injection of a variety of noxious agents, as well as trauma, hemorrhage, and purely emotional responses without overt physical damage.

It is therefore of considerable interest that we, as well as others, have shown that the injection of epinephrine in quantities within the physiological rate of secretion of the adrenal medulla is effective in reducing the number of circulating eosinophils or the level of adrenal ascorbic acid. Furthermore, by direct measurement in the adrenal vein blood of the dog, Vogt (1944) has shown that epinephrine produces an augmented secretion of cortical hormones.

Mode of Action of Epinephrine

There are several possibilities of explanation of the action of epinephrine in stimulating the adrenal cortical secretion. (1) By *directly* provoking the secretion by an effect on the adrenal cortical cells that is independent of the release of ACTH. (2) By *directly* stimulating the cells of the anterior pituitary responsible for the release of ACTH. (3) By activating a hypothalamic-anterior pituitary mechanism either by (*a*) the stimulation of adrenergic fibers passing directly to the anterior lobe or (*b*) by serving as the humoral component of a neuro-humoral link between the hypothalamic centers and the gland. (4) By increasing the "utilization" of cortical hormone in the tissues through its well-known capacity to increase the metabolic rate. Let us briefly consider each of these possibilities.

1. *Direct Effect on the Adrenal Cortex.* In our experience (Long, 1947) and that of others with the hypophysectomized rat, neither the injection of epinephrine nor exposure to conditions provoking epinephrine discharge brings about any fall in the circulating eosinophils or adrenal ascorbic acid. Our own results with totally hypophysectomized animals are so decisive that

we have concluded that epinephrine does *not* have a direct stimulating action on the adrenal cortex.

2. *Direct Action on the Anterior Pituitary.* By this is meant that epinephrine acts directly on the secretory cells of the anterior pituitary without the interposition of adrenergic fibers to the cells. This would be analogous to its capacity to accelerate glycogenolysis in skeletal muscle which is also believed to occur in the absence of such fibers.

It is evident that to draw a distinction between such an effect and the possible role of epinephrine as a humoral component of a neuro-humoral link is not easy to make so long as the effects of epinephrine are studied in animals in which the pituitary is left in connection with the central nervous system through its stalk.

3. *Hypothalamic-Hypophyseal Links.* The paucity of nerves reaching the anterior pituitary through the stalk is in sharp contrast to the abundance and demonstrated secretory function of those reaching the posterior lobe. This fact has made it difficult to support the view that the secretion of ACTH is dependent on nerve pathways from the hypothalamus which reach the cells of the anterior lobe through the pituitary stalk.

The fact that the gonadotropic function of the gland, at least in the rabbit, is markedly affected by nervous stimulation has led Harris (1947) to postulate the existence of a neuro-humoral link between the hypophysis and the gland. He has advanced the view that a humoral agent is liberated by the fibers reaching the tuberal region, and this agent is then transported by the portal system of blood vessels, described by Pope and Fielding, to the anterior lobe cells. The nature of the humoral agent is not known nor is it known whether the liberation of ACTH is inhibited by stalk section in which all possible reconnection by vascular means between the hypothalamus and pituitary body is prevented.

Recently in a short note Hume (1949) has suggested another type of neuro-humoral link governing ACTH secretion. He supposes that certain cells in the hypothalamus are stimulated through nerve pathways or by epinephrine to secrete a "hypothalamic hormone" which is then carried by the blood stream to the anterior lobe where it invokes the secretion of ACTH.

4. *Increase in the "Utilization" of Cortical Hormones by the Tissues.* As is well known, the injection of epinephrine brings about marked alternations in metabolism, including a significant increase in basal metabolic rate. It is therefore possible that the ensuing changes in the composition of the blood might act as an effective stimulus for ACTH secretion. Among such changes are a rise in blood glucose, lactic acid, and serum potassium. We have attempted to find whether increases in the blood of these substances in the magnitude encountered after epinephrine injection would cause depletion of adrenal ascorbic acid but have failed to demonstrate any significant effect. It is of course possible that epinephrine may induce changes in other blood constituents than those mentioned above and one of these may be the means by which the pituitary is stimulated. So far, however, the demonstration that the release of ACTH after the liberation of epinephrine is due to chemical changes on the blood traversing the gland is not forthcoming.

The fact that epinephrine increases the basal metabolic rate offers the possibility that its effects on ACTH secretion are due to an increased rate of "utilization" of cortical hormones by the tissues. Such an effect would be in keeping with Sayers' hypothesis, at least so far as the prolonged effect of epinephrine on cortical secretion is concerned. As in other instances, it would hardly seem to account for the rapid release of ACTH that follows the release or injection of the medullary hormone.

Experiments to Test the Above Alternatives

1. As stated before, we have no evidence that epinephrine can bring about the release of adrenal cortical hormones in the absence of the anterior lobe. This is true whether the eosinophil count or the level of adrenal ascorbic acid is used as a measure of such release.

2. Let us assume that the reflex secretion of epinephrine is the trigger mechanism that initiates the discharge of ACTH from the anterior pituitary. It follows, therefore, that any procedure that interrupts this reflex pathway should be followed by changes in the response that can be detected by the methods employed for the measurement of ACTH and ACH release. Of these, the level of circulating eosinophils is probably the most sensitive.

Using exposure to 4° C for 4 hours as the stimulus, we have shown that in normal rats there is a significant decline in 1 hour in circulating eosinophils which persists for at least 4 hours. However, when the reflex stimulation of epinephrine secretion is prevented (*a*) by removal of the adrenal medulla, (*b*) by spinal section at the level of the third or fourth dorsal vertebra, or (*c*) by appropriate diencephalic lesions, there is no decline in these cells in the first hour and frequently no fall even in 4 hours. Similar results are observed after the much milder stimulus imposed by the subcutaneous injection of normal saline. Nevertheless, the intravenous administration of epinephrine reproduces the normal response in either demedullated spinal animals or those with diencephalic lesions.

It appears, therefore, that destruction of the reflex arc for epinephrine secretion at any point prevents the immediate response of the anterior pituitary to conditions that normally are associated with its activation. Such interruption even prevents the immediate response to such a damaging procedure as laparotomy with intestinal manipulation, although it does not prevent the delayed fall (4 hours) in the eosinophils.

3. The observation that there is an immediate release of ACTH from the pituitary that is abolished by prevention of epinephrine release indicates the importance of the medullary hormone in the regulation of the adrenal cortical secretion. Nevertheless, the further observation that a *delayed* secretion can occur after removal of the adrenal medulla or interruption of the nerve pathway for epinephrine secretion clearly shows that this phase of the regulation of the anterior pituitary is not dependent on epinephrine release.

Since medullary stimulation can be excluded, the only explanation for the *delayed* response of the anterior pituitary-adrenal cortical mechanism would appear to be either (*a*) a lowering of the blood level of ACH due to the increased tissue "utilization" that may be associated with the circumstances or (*b*) the activation of a neural or neuro-humoral pathway leading to ACTH secretion from the pituitary. In an attempt to choose between these last two alternatives the following experiment was performed.

When normal animals are given a small subcutaneous injection of 10% sodium chloride, the ensuing painful stimulation causes a marked and persistent fall in blood eosinophils.

In rats whose spinal cords have been transected between the third and fourth dorsal vertebrae some weeks previously, similar injections *above the level of section* produce no fall, either immediate or delayed, in these cells even though the animals so injected exhibited the same degree of pain and discomfort as normal animals.

Two things should be noted in these experiments (1) that the stimulus is essentially one producing pain and a minimal amount of tissue damage. It is therefore unlikely that, in the absence of epinephrine release, any marked increase in metabolic rate would occur since the discomfort produced by the salt solution lasts only a few minutes. (2) In the spinal animal, although the nerve pathways for the reflex secretion of epinephrine are interrupted by the section, all sensory nerve pathways above the lesion are unaffected and presumably could cause reflex activation of any hypothalamic centers with which they were connected. Since such stimulation in normal animals does cause activation of ACTH secretion, it can only be presumed that in the absence of epinephrine discharge these neural or neuro-humoral pathways are ineffective in promoting ACTH release.

In consequence, our conclusion at the present time is that the regulatory mechanism for the secretion of ACTH, and consequently for adrenal cortical secretion, consists of at least two components.

The first of these is entirely humoral and is related to the relative blood content of ACTH and ACH. This relationship is an inverse one of a kind that a decline in the blood level of ACH is followed by ACTH secretion and conversely that increased levels of ACH inhibit ACTH release.

The second is a rapidly acting mechanism initiated by the release of epinephrine from the adrenal medulla. In the absence of such release the secretion of ACTH can occur by the humoral mechanism outlined above although at a much slower rate. Once epinephrine is released it appears to act rapidly by direct stimulation of the anterior lobe of the pituitary and more slowly by its capacity to increase the metabolic rate of the tissues.

Finally, our evidence at the present time does not support the view that in the absence of epinephrine neural or neuro-humoral mechanisms exist that can cause a rapid release of ACTH from the anterior lobe. At the moment the only hypothalamic centers and nerve pathways that appear to be associated with ACTH release in the normal animal are those that have already been identified as participating in the reflex secretion of epinephrine.

The experiments reported in this paper were carried out in collaboration with Doctors J. R. Brobeck, H. Gershberg, W. V. McDermott, Jr., and Miss E. G. Fry. We are indebted to the Fluid Research Fund of the Yale University School of Medicine for support of these studies.

REFERENCES

Harris, G. W., *J. Anat.*, **81**, 343 (1947).
Hume, D. M., *J. Clin. Invest.*, **28**, 799 (1949).
Long, C. N. H., *Federation Proc.*, **6**, 461 (1947).
Sayers, G., and Sayers, M., *Recent Progress in Hormone Research*, Vol. II., p. 81. Academic Press, New York, 1948.
Vogt, M., *J. Physiol.*, **103**, 317 (1944).

PHYSIOLOGICAL REGULATION OF THE ZONA GLOMERULOSA OF THE RAT'S ADRENAL CORTEX, AS REVEALED BY CYTOCHEMICAL OBSERVATIONS*

HELEN WENDLER DEANE
Harvard Medical School, Boston, Massachusetts

The present paper summarizes the cytochemical evidence suggesting that the zona glomerulosa of the rat's adrenal cortex is not under the influence of the anterior pituitary gland, but is under independent humoral control. The physiological conditions which alter the secretory activity of the glomerulosa suggest that its hormones affect chiefly the electrolyte balance of the body fluids and that the electrolytes, in turn, react upon the glomerulosa to affect its secretory rate.

The cells of the zona glomerulosa, like those of the fasciculata, are filled with small fatty droplets; the two zones are separated by a few rows of relatively fat-free cells (Fig. 3; see also Figs. 5 and 22, Deane and Greep, 1946). The fatty droplets may be distinguished from other lipids by a number of characteristic reactions. (1) In unfixed sections of the gland, the droplets display a Maltese cross when viewed under the polarizing microscope and, after fixation in formalin, they contain birefringent crystals (Fig. 13, Olson and Deane, 1949). (2) They develop a blue-green color when treated with a mixture of sulfuric and acetic acids after oxidation with ferric ammonium sulfate (the Schultz reaction) (Nichols, 1948). (3) After formalin fixation, they emit a yellowish or greenish fluorescence when examined in ultraviolet light (Fig. 8, Deane and Greep). (4) They react with various carbonyl reagents, such as phenylhydrazine, the hydrazide of hydroxynaphthoic acid (Ashbel and Seligman, 1949) (Fig. 2), and Schiff's leucofuchsin without pretreatment with mercuric chloride. (5) All these characteristics are obliterated if the sections are extracted beforehand with acetone at room temperature. Taken together, this group of tests indicates the presence of lipids in which the carbonyl groups, optical activity and unsaturation are similar to those of ketosteroids. Hence it has been suggested that they localize the sites of formation of the steroid hormones (Dempsey, 1948; Greep and Deane, 1949), although none of the reactions is specific for these compounds.

Regardless of the specificity of the tests, the number, size, and reactivity of the droplets change during induced activity or inactivity of the gland. Consequently, we need not decide here whether these substances are the hormones and their precursors or only metabolically related compounds. The

* This work was aided by grants from the American Cancer Society (1) on the recommendation of the Committee on Growth of the National Research Council and (2) from an Institutional Grant to Harvard University.

overall cytological changes in the cells can be used as an index of secretory activity. In both the glomerulosa and the fasciculata, the cells multiply and enlarge when stimulated (Fig. 1, *A*→*B*-*D*). Their droplets become small and, with a moderate stress, increase in number (*A*→*C*). If the stress is more severe, they may disappear (*A*→*B*, or *A*→*C*→*D*). On the other hand, when the cells are unstimulated, they shrink (*A*→*E*-*G*). The droplets

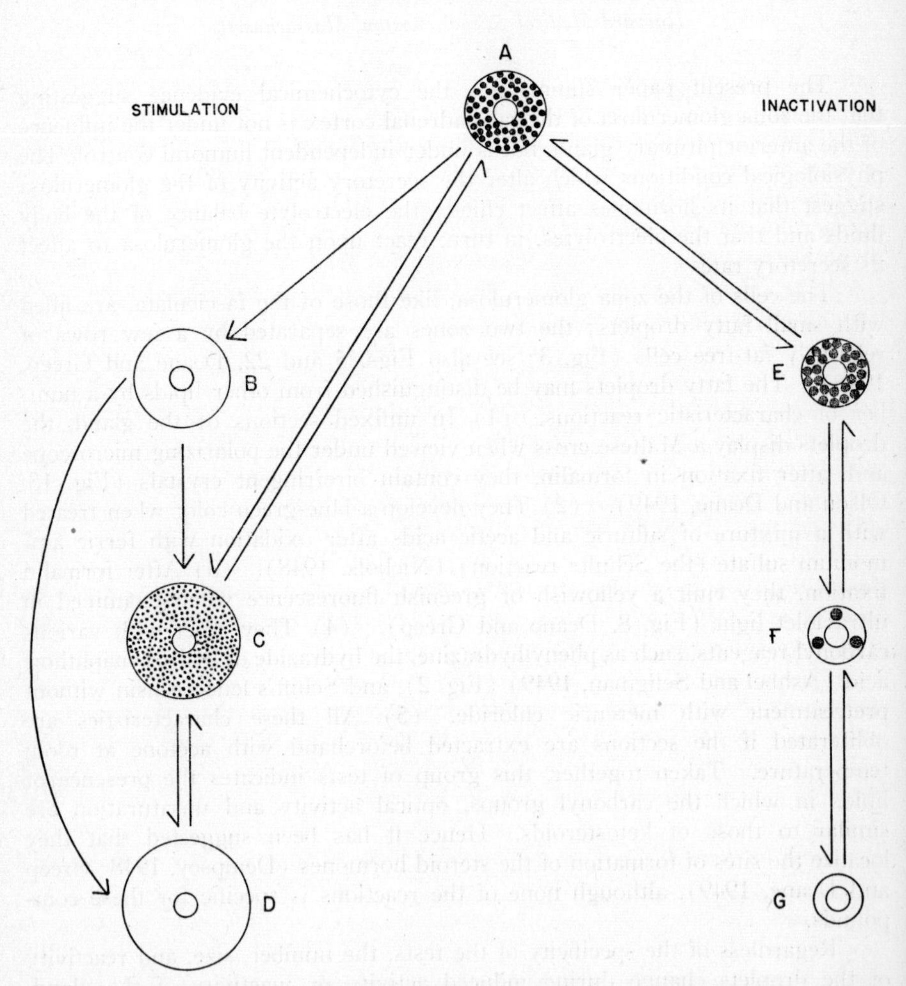

FIG. 1. Modulation of adrenal cortical cells in either the glomerulosa or in the fasciculata under conditions of hyperactivity (*B* to *D*) or of underactivity (*E* to *F*). When stimulated, the cells, and hence the zone in which they lie, hypertrophy. The lipid droplets become smaller than normal and may, under extreme conditions, disappear. When stimulus is removed, the cells atrophy. The lipid droplets at first enlarge and decrease in number; later they may disappear. [From original by Deane, Shaw, and Greep, *Endocrinology,* **43,** 133 (1948).]

at first enlarge in size but decrease in number; later they may disappear. Comparable changes in cellular appearance and in secretory antecedents characterize activity or inactivity in other glandular cells, for example, those of the anterior pituitary gland (Wolfe, 1949). The subsequent interpretations of secretory activity of the zona glomerulosa are based, therefore, on the appearance of the cells as well as on that of their lipid droplets.

The Relation of the Pituitary Gland and the Glomerulosa

After hypophysectomy of the rat, the adrenal cortex shrinks to about one-half its normal cross-sectional area in 2 weeks and decreases more slowly thereafter. This shrinkage results from the atrophy and disappearance of cells lying in the fasciculata and reticularis. The glomerulosa actually increases in width. This increase depends in part on the maintenance of the volume of the zone on the surface of the shrinking sphere. The cells of the glomerulosa become somewhat smaller after hypophysectomy, however, and hyperplasia contributes to the maintenance of volume.

The cells of the inner zones gradually lose all their reactive lipids. On the other hand, the droplets in the cells of the glomerulosa retain their characteristic physical and chemical properties for at least 4 months, becoming, if anything, more reactive. Since hypophysectomized rats do not exhibit a fatal electrolyte imbalance (as do adrenalectomized rats), it appears that salt-regulating hormones are produced by the adrenal in the absence of the pituitary gland. The persistence of the glomerulosa after hypophysectomy strongly suggests that this zone is the source of the salt-regulating principles (Deane and Greep). This hypothesis was originally formulated by Swann (1940). Additional evidence for the continued activity of the zona glomerulosa after hypophysectomy has come from persistence of ascorbic acid (Deane and Morse, 1948) and alkaline phosphatase (Dempsey, Greep, and Deane, 1949) in this zone.

The known effects of posterior pituitary extracts on fluid and electrolyte balance suggest that the removal of this gland might affect the glomerulosa after total hypophysectomy. Dr. Greep and I have made some observations on the adrenals of rats from which the posterior pituitary alone was removed. The animals were killed at periods up to 5 months after the operation. Following an initial loss, these rats later gained weight. Marked polyuria persisted in some of them for the entire period. The adrenals were in general of normal size for 200-g male rats (*ca.* 15 mg per 100 g body weight). In those rats killed up to 2 months after the operation, the zona glomerulosa appears essentially normal in width and in lipid content. In the three killed later, however, the glomerulosa is somewhat broader than normal (Fig. 7). Some of the enlargement of this zone in hypophysectomized rats, therefore, may result from the reduction of posterior pituitary hormones.

The converse situation to hypophysectomy occurs when the secretion of the adrenocorticotropic hormone is increased. When the pituitary is stimulated by environmental stress, the zona fasciculata enlarges and appears hyperactive, but the glomerulosa is unaffected (Deane and McKibbin, 1946;

Deane and Shaw, 1947). Likewise, the administration of adrenocorticotropin fails to induce signs of increased secretory activity in the outer zone or to cause sodium retention in the rat (Bergner and Deane, 1948).

Humoral Changes Causing Signs of Decreased Secretion by the Glomerulosa

The steroid with the most conspicuous salt-regulating activity is desoxycorticosterone. Although the physiological occurrence of this particular compound is questionable, compounds possessing salt-retaining activity are present in adrenal extracts and can be distinguished by bioassay from those concerned primarily with organic metabolism (Olson, Jacobs, *et al.,* 1944). Consequently, desoxycorticosterone acetate may be used as a model substance until we know the precise chemical nature of those native hormones with this function. When daily injections of this steroid are given to the rat, the cells of the glomerulosa atrophy. The reactive droplets enlarge and gradually disappear until, after a month, the cells are devoid of visible lipid (Fig. 4). The response occurs in hypophysectomized as well as in intact rats. This phenomenon indicates that the injected steroid substitutes for the natural products of the glomerulosa and has been termed "disuse atrophy" (Greep and Deane, 1947).

Desoxycorticosterone and adrenal extracts cause not only the retention of sodium but also the excretion of potassium (Feil and Dorfman, 1945) and therefore elevate the sodium-potassium ratio in the blood stream. Atrophy of the glomerulosal cells and enlargement of the lipid droplets occur when the rat is fed a diet containing virtually no potassium (compare Figs. 3 and 5) (Nichols, 1948; Deane, Shaw, and Greep, 1948). Such a diet produces an elevated sodium-potassium ratio which resembles the effect of desoxycorticosterone. The hormone requirement may thus be reduced, and an explanation is afforded for the disuse atrophy observed in the glomerulosa. In 1949, Knowlton, Loeb, Seegal, and Stoerk reported shrinkage of the glomerulosa in rats with an increased intake of sodium chloride, a treatment which would also elevate the sodium-potassium ratio. We have some observations which substantiate their findings (Deane and Masson, in press).

It is perhaps of interest to report that no glomerulosal alteration occurs when the calcium-magnesium ratio is increased by feeding rats a diet deficient in magnesium (Lytt Gardner and Deane, unpublished). The response thus seems limited to raising the ratio of the monovalent cations.

Humoral Changes Causing Signs of Increased Secretion by the Glomerulosa

In contrast to the experimental situations mentioned in the preceding section, the feeding of a diet lacking sodium lowers the sodium-potassium ratio in the blood. After such treatment, the cells of the glomerulosa enlarge, and the lipid droplets become small and eventually disappear (compare Figs. 3 and 8). The increased secretory activity probably results from the stimulus of the lowered sodium-potassium ratio (Nichols; Deane, Shaw, and Greep). Related to this observation, Leaf, Couter, *et al.* (1949) obtained evidence in

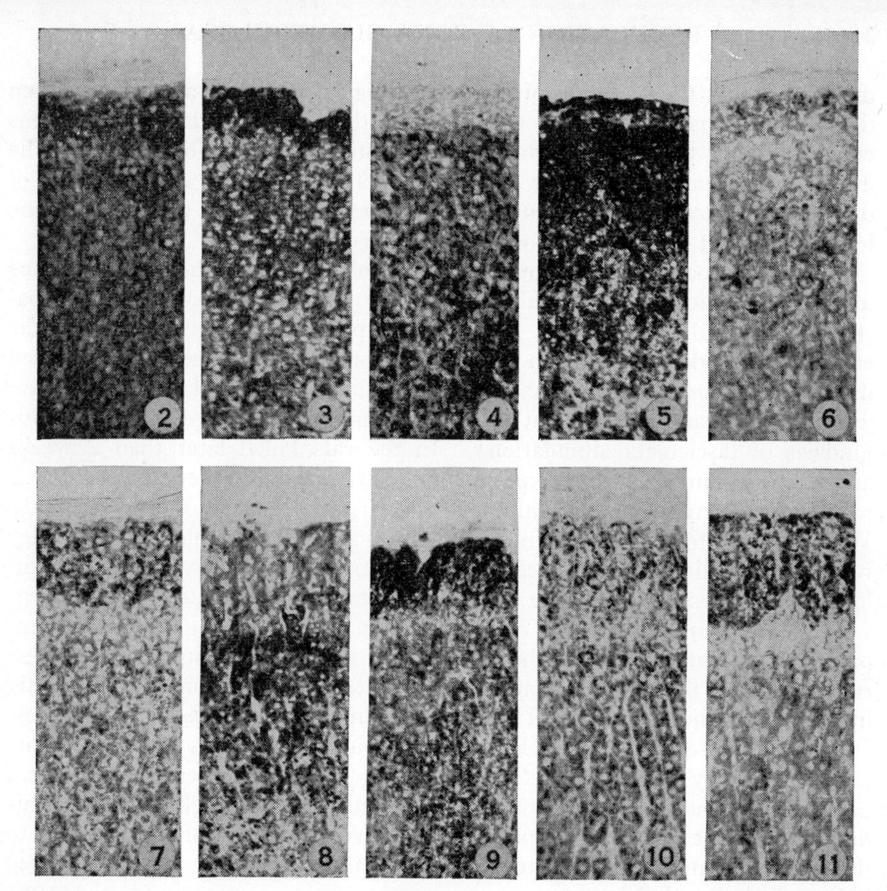

Adrenal glands which were fixed in formalin and sectioned on the freezing microtome at 15 μ. Figures 2-11 were stained with sudan black B. All × 160.

FIG. 2. Gland from normal rat. Stained by the Ashbel-Seligman method for ketonic steroids. The lipid droplets in the cortex are stained blue-purple by this reaction.

FIG. 3. Gland from rat which received a balanced purified diet, showing the normal width and lipid content of the glomerulosa. Control for Figs. 5, 8, and 9. (From original by Deane, Shaw and Greep, *Endocrinology*, **43**, 133.)

FIG. 4. Gland from rat which received a 2.5-mg desoxycorticosterone acetate daily for 111 days. (Blood pressure, 190 mm Hg.) The glomerulosa is shrunken and devoid of lipid (Deane and Masson, in press).

FIG. 5. Gland from rat which received a potassium-free diet for 70 days. The glomerulosa is shrunken but still contains large lipid droplets. (From original by Deane, Shaw, and Greep, *Endocrinology*, **43**, 133.)

FIG. 6. Gland from rat which was unilaterally nephrectomized for 41 days. The glomerulosa is of normal width and contains a moderate amount of lipid. This serves as a control for Fig. 11 (Deane and Masson, in press).

FIG. 7. Gland from rat which had the posterior pituitary removed 5 months previously. The glomerulosa is slightly enlarged but not depleted of lipid (Greep and Deane, unpublished).

FIG. 8. Gland from rat which received a sodium-free diet for 22 days. The glomerulosa is broad and virtually depleted of lipid (Deane, Shaw, and Greep).

FIG. 9. Gland from rat which received a diet lacking both sodium and potassium for 38 days. The glomerulosa is broad but contains a large amount of lipid. (From original by Deane, Shaw, and Greep, *Endocrinology*, **43**, 133.)

FIG. 10. Gland from a rat which received a chloride-deficient diet for 5 weeks. The glomerulosa is broadened but not depleted (Locke and Deane, unpublishd).

FIG. 11. Gland from a rat which had both kidneys encapsulated for 73 days. (Blood pressure, 180 mm Hg.) The glomerulosa is broad as compared with its control (Fig. 6) but contains an increased number of lipid droplets (Deane and Masson, in press).

the human that the secretion of salt-regulating hormones is augmented when the diet contains little sodium chloride. In the rat, two other interventions which decrease the sodium-potassium level stimulate the glomerulosa. The injection of potassium chloride induces a rapid enlargement and depletion of the zone. A moderate enlargement with no depletion of lipid characterizes the feeding of a diet lacking both sodium and potassium (Fig. 9).

All the procedures mentioned above involve shifts in the balance of the monovalent cations. In the fall of 1949, Dr. William Locke of the Massachusetts General Hospital and I made some preliminary observations on the effects of chloride deficiency on the adrenal cortex. Rats were fed a purified diet in which phosphate was substituted for chloride in the salt mixture. There was no significant enlargement of the adrenals or shrinkage of the thymus (indices of fasciculata stimulation). In the rats killed later than 2 weeks after the beginning of the experimental diet, however, the glomerulosa is wider than normal (Fig. 10), although there is no depletion of reactive lipids. We conclude tentatively from these observations that chloride deficiency results, either directly or indirectly, in a moderate stimulation of the glomerulosa. Further experiments are planned to obtain an explanation for this.

Kidney damage in acute choline deficiency of weanling rats is accompanied by an enlargement of the glomerulosa and a loss of its lipids. These rats experience marked renal failure, with a decreased sodium-potassium ratio in the blood and a high level of non-protein nitrogen (Olson and Deane, 1949). Because of the high NPN, we examined the adrenals from a few rats made uremic by removing the kidneys or by ligating the renal arteries and ureters. In none of these was there any stimulation of the glomerulosa. This finding does not exclude the possibility, however, that part of the stimulus to the outer zone may have resulted from a metabolite released by the damaged kidney of choline deficiency.

To investigate the suggestion that damaged kidneys may release a product which stimulates the glomerulosa, I obtained from Dr. Georges Masson of the Cleveland Clinic adrenal glands from rats in which the kidneys were encapsulated for long periods. Such rats display marked hypertension. In these, also, the glomerulosa shows an increase in width, an enlargement of its cells, and an increased lipid content (compare Figs. 6 and 11). The birefringent crystals appear uniformly fine, rather than mixed in size as they are in glands of normal rats. The response is not the hypertension *per se,* since animals made hypertensive by desoxycorticosterone (Fig. 4) or by an anterior pituitary preparation (Masson, Corcoran, and Page, 1949) fail to show enlargement of the glomerulosa. Nor is there any significant alteration in blood electrolytes in rats made hypertensive by constriction of the kidney. If anything, there is a slight retention of sodium and loss of potassium (Friedman, 1948). Further experiments are planned to determine the identity of the stimulating factor (Deane and Masson).

Comment

These data indicate that the zona glomerulosa, in the rat, secretes hormones that influence the retention of sodium (and possibly of chloride) and the excretion of potassium. Secretion by this zone apparently continues in the absence of the pituitary and is not affected by the administration of adrenocorticotropin. The secretory activity of the glomerulosa seems, therefore, to be independent of the anterior pituitary. Despite this independence, the metabolism of the glomerulosa may be depressed by hypophysectomy. Joseph, Schweizer, Ulmer, and Gaunt (1944) found that forced water intake induces a more rapid diuresis in normal than in hypophysectomized rats. In the latter animals, the abnormality is partially repaired by the administration of desoxycorticosterone. Thus, although the glomerulosa of the hypophysectomized rat may function normally under ordinary laboratory conditions, its secretory reserve may be impaired and this impairment may be revealed by a sudden and maximal stress. Regardless of the reserve potential of the gland, however, its ordinary fluctuations in secretory activity would seem to be controlled by the electrolyte balance of the blood, or by metabolites from the kidney, and not by the anterior pituitary gland.

The material discussed so far has pertained entirely to the rat. In man, however, adrenocorticotropin induces salt retention. This fact has been taken to indicate that the human glomerulosa may be regulated by the anterior pituitary gland. The question arises, therefore, whether the apparent independence of the rat's glomerulosa is a species peculiarity or whether the phenomenon is a general one which may be demonstrated in other species.

The glomerulosa remains normal in appearance after hypophysectomy in several species other than the rat, for example, the mouse (Jones, 1949), the guinea pig (Schweizer and Long, 1950) and the dog (Houssay and Sammartino, 1933). Like the rat, these species do not succumb in severe electrolyte imbalance following hypophysectomy. Moreover, it is the clinical impression that hypo-pituitarism in the human does not result in the marked inability to regulate salt excretion that characterizes the Addisonian patient, although this is not an established fact (Sheehan and Summers, 1949). Thus, it would appear to be a tenable hypothesis that the glomerulosa may function without pituitary stimulation in many species, possibly even in man. Although this hypothesis is as yet unproved for man, it seems that the difference between man and the rat may be one of degree rather than kind.

REFERENCES

Ashbel, R., and Seligman, A. M., A new reagent for the histochemical demonstration of active carbonyl groups. A new method for staining ketonic steroids. *Endocrinology,* **44,** 565 (1949).

Bergner, G. E., and Deane, H. W., Effects of pituitary adrenocorticotropic hormone on the adult rat, with special reference to cytochemical changes in the adrenal cortex. *Endocrinology,* **43,** 240 (1948).

Deane, H. W., and Greep, R. O., A morphological and cytochemical study of the rat's adrenal cortex after hypophysectomy, with comments on the liver. *Am. J. Anat.,* **70,** 117 (1946).

Deane, H. W., and Masson, G. M. C., Adrenal cortical changes in rats with various types of experimental hypertension. *J. Clin. Endocrinol.* (in press).

Deane, H. W., and McKibbin, J. M., The chemical cytology of the rat's adrenal cortex in pantothenic acid deficiency. *Endocrinology,* **38,** 385 (1946).

Deane, H. W., and Morse, A., The cytological distribution of ascorbic acid in the adrenal cortex of the rat under normal and experimental conditions. *Anat. Record,* **100,** 127 (1948).

Deane, H. W., and Shaw, J. H., A cytochemical study of the responses of the adrenal cortex of the rat to thiamine, riboflavin, and pyridoxine deficiencies. *J. Nutrition,* **34,** 1 (1947).

Deane, H. W., Shaw, J. H., and Greep, R. O., The effect of altered sodium and potassium intake on the width and cytochemistry of the zona glomerulosa of the rat's adrenal cortex. *Endocrinology,* **43,** 133 (1948).

Dempsey, E. W., "The Chemical Cytology of Endocrine Glands," in *Recent Progress in Hormone Research,* Vol. 3, p. 127. Academic Press, New York, 1948.

Dempsey, E. W., Greep, R. O., and Deane, H. W., Changes in the distribution and concentration of alkaline phosphatases in tissues of the rat after hypophysectomy or gonadectomy, and after replacement therapy. *Endocrinology,* **44,** 88 (1949).

Feil, M. L., and Dorfman, R. I., The influence of urinary cortin-like material on sodium and potassium metabolism. *Endocrinology,* **37,** 437 (1945).

Friedman, S., Plasma electrolytes in hypertension produced by renal compression. *Rev. can. biol.,* **7,** 570 (1948).

Greep, R. O., and Deane, H. W., Cytochemical evidence for the cessation of hormone production in the zona glomerulosa of the rat's adrenal cortex after prolonged treatment with desoxycorticosterone acetate. *Endocrinology,* **40,** 417 (1947).

Greep, R. O., and Deane, H. W., The cytology and cytochemistry of the adrenal cortex. *Ann. N. Y. Acad. Sci.,* **50,** 596 (1949).

Houssay, B. A., and Sammartino, R., Modifications histologique de la surrénale chez les chiens hypophysoprives ou a tuber lésé. *Compt. rend. soc. biol.,* **114,** 717 (1933).

Jones, I. Chester, The relationship of the mouse adrenal cortex to the pituitary. *Endocrinology,* **45,** 514 (1949).

Joseph, S., Schweizer, M., Ulmer, N. Z., and Gaunt, R., The anterior pituitary and its relation to the adrenal cortex in water diuresis. *Endocrinology,* **35,** 346 (1944).

Knowlton, A., Loeb, E. N., Seegal, B. C., and Stoerk, H. C., Desoxycorticosterone acetate: Studies on the reversibility of its effect on blood pressure and renal damage in rats. *Endocrinology,* **45,** 435 (1949).

Leaf, A., Couter, W. T., Lutchonsky, M., and Reimer, A., Evidence that renal sodium excretion by normal human subjects is regulated by adrenal cortical activity. *J. Clin. Invest.,* **28,** 1067 (1949).

Masson, G. M. C., Corcoran, A. C., and Page, I. H., Experimental vascular diseases due to desoxycorticosterone acetate and anterior pituitary extract. I. Comparison of functional changes. *J. Lab. Clin. Med.,* **34,** 1416 (1949).

Nichols, J., Effects of electrolyte imbalance on the adrenal gland. *Arch. Path.,* **45,** 717 (1948).

Olson, R. E., and Deane, H. W., A physiological and cytochemical study of the kidney and the adrenal cortex during acute choline deficiency in weanling rats. *J. Nutrition,* **39,** 31 (1949).

Olson, R. E., Jacobs, F. A., Richert, D., Thayer, S. A., Kopp, L. J., and Wade, N. J., The comparative bioassay of several extracts of the adrenal cortex in tests employing four separate physiological responses. *Endocrinology,* **35,** 430 (1944).

Schweizer, M., and Long, M. E., Experimental studies of the anterior pituitary: Partial maintenance of the adrenal cortex by anterior pituitary grafts in fed and starved guinea pigs. *Endocrinology,* **46,** 191 (1950).

Sheehan, H. L., and Summers, K. V., The syndrome of hypopituitarism. *Quart. J. Med.,* **18,** 319 (1949).

Swann, H. G., The pituitary-adrenocortical relationships. *Physiol. Rev.,* **20,** 493 (1940).

Wolfe, J. M., Cytochemical studies on the anterior hypophysis of rats receiving estrogen. *Am. J. Anat.,* **85,** 309 (1949).

ON CERTAIN FACTORS CONDITIONING THE ACTION OF THE PITUITARY-ADRENAL SYSTEM

FLOYD R. SKELTON

Institute of Experimental Medicine and Surgery, University of Montreal, Montreal, Canada

Since first formulated in 1936 (Selye, 1936), the concept of a systemic response of an organism subjected to nonspecific stress in which the activation of the pituitary-adrenal system is of primary importance has steadily gained in both physiological and pathological significance. Today we are aware that the pituitary-adrenal system can become activated by a vast number of stimuli and that a series of morphological and biochemical changes ensue which have been described as the general-adaptation-syndrome (Selye, 1950). Furthermore, there is a considerable amount of evidence to support the contention that perversion of this defense mechanism causes profound detrimental effects which become manifest in pathological lesions. Although the bulk of research has been turned to the study of both the physiological and pathological changes produced by the hormones of the pituitary-adrenal system, we feel that it is of equal importance to know how the terrain or internal environment in which these hormones function may affect their actions. There is a small but growing literature which suggests that through experimentally produced variations in the internal metabolic environment, the functional activity of the pituitary-adrenal system as well as the effects of the hormones of this system can be modified or conditioned.

The purpose of this paper, therefore, is twofold: First, to review those observations which indicate that the activity of the pituitary-adrenal system can be modified by factors in the internal environment. Since space is limited only those factors which are known to act at the pituitary, adrenal, and peripheral levels respectively will be considered. Second, to discuss the significance of this concept in relation to the pathogenesis of a group of human diseases which Selye has called the Diseases of Adaptation.

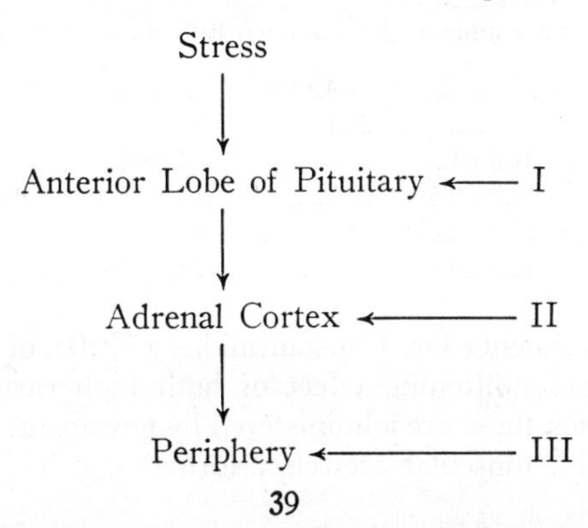

The above illustration indicates schematically the sites at which conditioning will be discussed: I Pituitary, II Adrenal, III Periphery (post-adrenal level).

Conditioning at the Pituitary Level

During the initial investigations of the diseases of adaptation it was observed that the proportion of protein in the diet markedly altered the experimental results (Berman *et al.*, 1945). Subsequently it was found that rats subjected to prolonged stress or treatment with a crude anterior pituitary preparation (LAP) and maintained on a 30% protein diet developed cardiovascular changes, marked nephrosclerosis, adrenal hypertrophy, and hypertension. On the other hand, rats maintained on a 15% protein diet either failed to develop these pathological changes or did so to a much less extent (Dontigny *et al.*, 1948; Hay *et al.*, 1948).

Further studies concerning the conditioning of pituitary-adrenal function by dietary protein have been conducted by Moya *et al.* (1948). They found that rats exposed to cold for 1 hour following pretreatment with a 30% protein ration for a period of two weeks exhibited a significantly greater depletion of adrenal ascorbic acid than rats pretreated with a 15% protein diet. Likewise these workers have shown that the compensatory adrenal hypertrophy which follows unilateral adrenalectomy is significantly greater when the rats are fed a 30% protein ration than it is following the administration of a 15% protein diet. Although all investigators do not agree with these effects of dietary protein on adrenal reactivity (Ingle *et al.*, 1943; Benna and Harvard, 1945), recent results obtained by Henriques, Henriques, and Selye (1949) indicate that the adrenal enlargement and ascorbic acid depletion which follow LAP treatment are increased by a high protein intake. From these experiments it has been concluded that the degree of reactivity of the pituitary-adrenal system is significantly affected by the dietary protein concentration.

TABLE 1

Effect of High and Low Amino Acid Diets on the Adrenal Weight Response to Various Alarming Stimuli

	Controls			Experimentals	
	5% amigen	25% amigen	40% amigen	5% amigen	40% amigen
Cold	....	17.4±0.6	....	20.2±1.2	25.3±1.1
Formaline I	....	23.1±1.0	....	24.4±1.3	27.5±1.2
Formaline II	21.0±1.3	....	22.0±1.0	22.2±0.7	24.9±1.2
Exercise	12.6±0.5	....	14.3±0.4	14.4±0.6	19.7±1.3
Urethane	15.8±0.8	....	17.2±0.5	18.5±1.2	19.8±0.7
Colchicine	12.6±0.5	....	14.4±0.4	19.1±0.7	16.2±1.1

In recent experiments Dr. Constantinides (1950) of our laboratory has studied in rats the conditioning effect of both high carbohydrate and high amino acid diets when these are administered by gavage during stress produced by exposure to cold, muscular exercise, formalin, colchicine, and urethane.

The results of this study, with the exception of colchicine, have indicated that the adrenal hypertrophy, lymphatic tissue breakdown, and liver glycogen depletion are minimal on a high carbohydrate, low amino acid intake and maximal on a diet of the opposite composition. Table 1 illustrates the effects of these diets on the adrenal weight. It can be seen that in general the greatest hypertrophy occurred in the rats receiving the high amino acid intake. Since colchicine stress was affected in exactly the reverse direction it is suggested that all stresses are not conditioned in the same manner, a fact which may be related to the different metabolic demands made upon the organism by various agents.

Since the experiments thus far discussed had been done in intact animals they could not reveal at what level of the pituitary-adrenal system the dietary protein exerted its activity. To investigate this point Moya *et al.* (1948) studied the effect of high and low protein intake on the adrenal response to ACTH in hypophysectomized rats. Using the decrease of adrenal ascorbic acid as the criterion of increased functional activity consequent to ACTH administration, these investigators found that the ascorbic acid depletion was identical whether the animals had been prefed a 30 or a 15% protein diet. These results have also been confirmed by Henriques *et al.* (1949) who found the adrenal ascorbic acid depletion following LAP administration in the hypophysectomized rat to be unaffected by the dietary protein concentration. Thus, since the protein of the diet did not alter the action of ACTH or LAP on the adrenal, it seemed by inference that its site of action was at a higher level, probably the pituitary.

From all the foregoing data it seems justified to conclude that if the protein content of the diet is increased at the expense of the carbohydrate content, the reactivity of the pituitary-adrenal system to some stresses can be increased whether the stress be of short or long duration. Furthermore, it seems that the dietary protein exerts its conditioning action in these instances at the pituitary level and probably functions by providing the amino acid building blocks necessary for the synthesis of increased amounts of ACTH.

Conditioning at the Adrenal Level

It has previously been reported that glucose administration was capable of modifying some of the criteria of adrenal activity in the intact animal (Abelin, 1945; Steeples and Jensen, 1949). To investigate this possibility further and to obtain information as to the site of action of this factor, the following preliminary experiment was performed in association with Drs. P. Constantinides and C. Fortier. A single injection of 4.7 mg of Armour's ACTH (Batch No. 64-A) was given to rats 24 hours after hypophysectomy. Two and one-half hours after the ACTH injection, 1 g of glucose was administered by stomach tube. At intervals of 30 minutes, 1, 3, and 6 hours after the glucose administration, the levels of adrenal cholesterol and ascorbic acid as well as of blood glucose were studied. It was found that the ACTH-induced adrenal cholesterol depletion was enhanced by the administration of glucose while the pattern of adrenal ascorbic acid response to ACTH was not

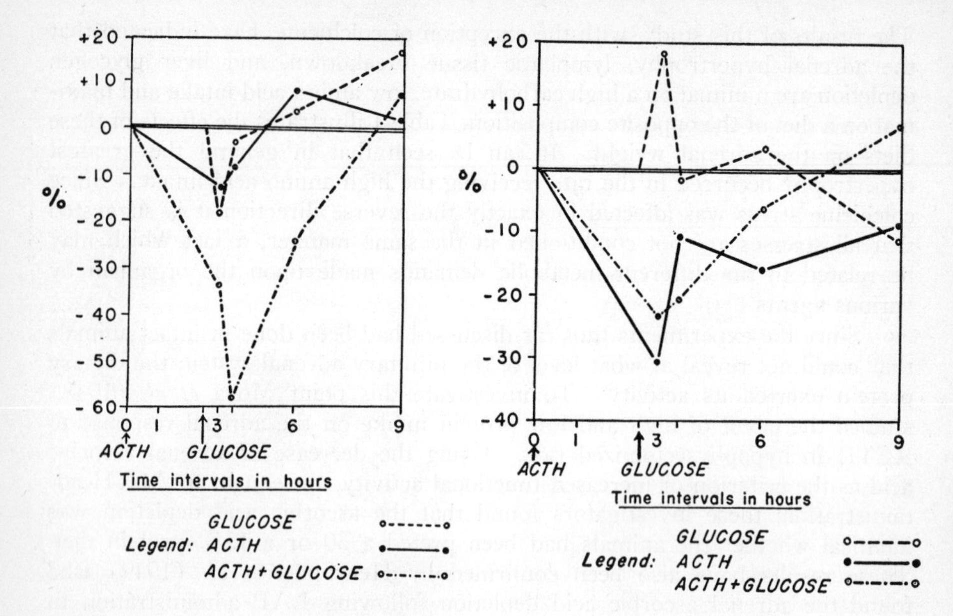

FIG. 1. Effect of oral glucose administration on the adrenal cholesterol response to ACTH in the hypophysectomized animal —percentage of deviation from initial level. Although the lines representing both the ACTH- and ACTH plus glucose-treated groups are drawn for clarity from zero time as shown, it is to be understood that the adrenal cholesterol depletion of the latter group is the same as the former until glucose is given. The precipitous fall which occurs after the glucose administration and the slower return to normal levels thereafter are evident.

FIG. 2. Effect of oral glucose administration on the adrenal ascorbic acid response to ACTH in the hypophysectomized animal—percentage of deviation from initial level. Although the lines representing both the ACTH- and ACTH plus glucose-treated groups are drawn for clarity from zero time as shown, it is to be understood that the decrease in adrenal ascorbic acid in the latter group is the same as the former until glucose is given. As can be seen, the general pattern of ascorbic acid response is unaltered by glucose except for the suggestion of a more rapid return to normal levels.

altered (Figs. 1 and 2). However, this divergent effect on these two metabolites is probably explained by the fact that 3½ hours after ACTH treatment does not correspond to the maximum fall of adrenal ascorbic acid while it does for adrenal cholesterol. Subsequent experiments have now shown that neither insulin hypoglycemia nor adrenaline hyperglycemia influences the normal response of the adrenal cholesterol and ascorbic acid to administered ACTH. If, as these preliminary data seem to indicate, glucose can modify the response of the adrenal to exogenous ACTH, it is in some way connected with the increase in carbohydrate stores available to the organism.

A recent experiment performed in association with Dr. C. Fortier has shown that pretreatment with corticotropin inhibits the loss of cortical lipid granules which is normally produced in the rat by folliculoid hormones

(Skelton *et al.*, 1949) or stress (Selye and Stone, 1950) while the depletion of adrenal cholesterol and ascorbic acid is enhanced. Furthermore, the chronic administration of a crude anterior pituitary preparation (LAP) in the presence of necrosis has caused myeloid metaplasia of adrenal cortical cells with the formation of both erythropoietic and leukopoietic elements. Since testoid compounds such as methyl testosterone cause the transformation of adrenal cortical cells into ordinary fat cells the simultaneous administration of LAP and methyl testosterone produces hematopoiesis in a fat cell-containing stroma, thus imitating bone-marrow tissue. These experiments demonstrate that the hormones produced by other members of the endocrine system can exert a marked conditioning action on the function of the pituitary-adrenal axis. We shall see later that hormones may also act in a conditioning capacity at the peripheral (post-adrenal) level.

In this context it is interesting that Ralli (Dumm, Ovando, Roth, and Ralli, 1949) has shown that the pantothenic acid-deficient rat does not respond to ACTH with the usual lymphopenia. This suggests a nonresponsive or exhausted state of the adrenal. We found, that under ideal conditions the adrenals of rats deficient in this vitamin became hemorrhagic only after exposure to stress (Skelton, 1950), while the various peripheral signs of increased cortical hormone secretion, such as thymus atrophy, did not occur or were less pronounced than in either normal or paired-fed controls. Thus pantothenic acid deficiency seemingly decreases the responsiveness of the adrenal to either endogenous or exogenous ACTH.

On the other hand Giroud and his co-workers have repeatedly observed (Giroud *et al.*, 1940; 1941) that ascorbic acid administration favors the production of adrenal cortical hormones, a conditioning effect which undoubtedly acts at the adrenal level. Further studies of the relationship between ascorbic acid and adrenal-cortex function have been performed by Dugal and Thérien (Dugal and Thérien, 1949) who recently reported that the administration of large doses of this vitamin in the rat and guinea pig completely prevented the adrenal hypertrophy normally resulting from chronic exposure to cold. Since these experiments were done in intact animals, it is impossible to establish the site of this conditioning action but at least the modified pituitary-adrenal activity was manifest at the adrenal level.

Conditioning at the Peripheral (Post-Adrenal) Level

The hypophysectomized or adrenalectomized rat, which is notoriously sensitive to damage, responds to stress with a hypoglycemia. If, however, a small dose of adrenal cortical extract, which in itself barely alters the blood sugar level, is administered before the onset of stress, such rats respond with a hyper- rather than a hypoglycemia. This finding published by Selye and Dosne (1941) was one of the first observations made in this laboratory which indicated that the actions of cortical hormones could be conditioned. It was presumed that the altered metabolic conditions consequent to stress created a more ideal milieu for cortical hormone function which in this instance was manifest as hyperglycemia secondary to increased gluconeogenesis.

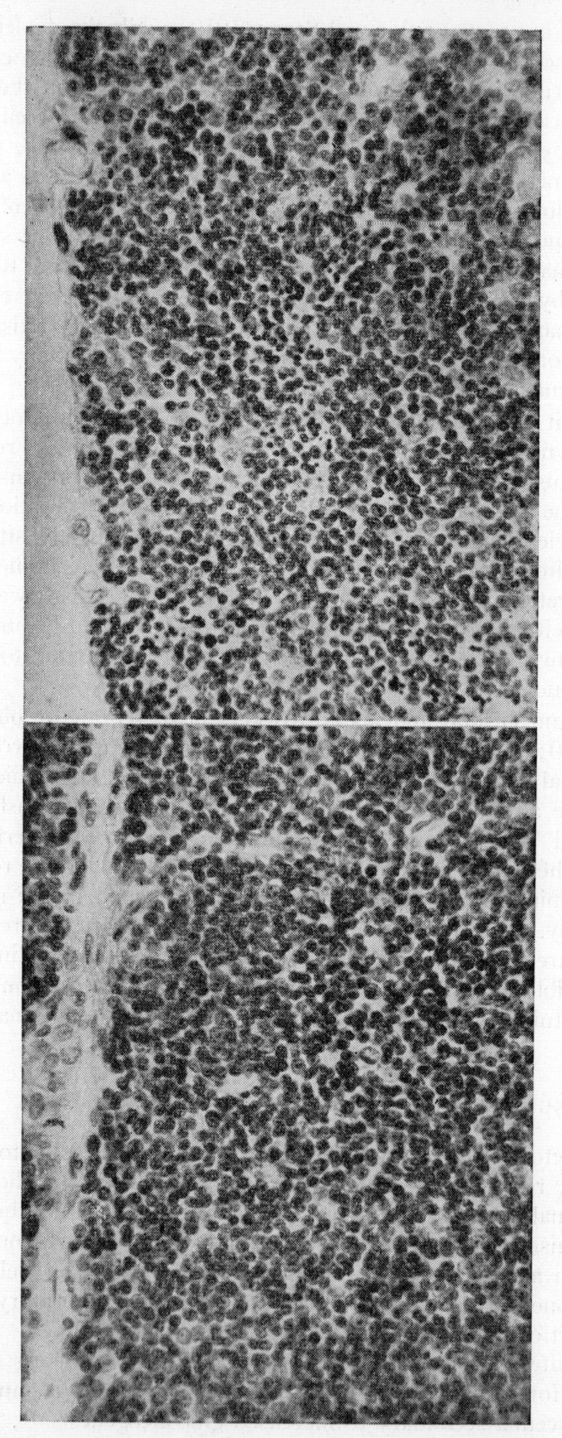

Fig. 3. A hematoxylin and eosin section of the thymus of an adrenalectomized rat treated with 0.2 ml of lipo-adrenal cortex extract (Upjohn). The normal appearance of the cellular structure with absence of caryoclasia of the thymocytes can be clearly seen.

Fig. 4. A similar section of thymus from a rat treated with the same dose of cortical extract but exposed to muscular exercise. Note the development of confluent areas of cary-oclasia of the lymphatic elements.

Comparative studies of the peripheral effects of stress and ACTH done in conjunction with other members of the Institute of Experimental Medicine and Surgery of the University of Montreal (Fortier *et al.*, 1950) have suggested that ACTH administration does not duplicate stress in all respects. Again the difference between the two conditions lay in the metabolic changes incident to stress itself. To investigate further the peripheral conditioning effects of stress-induced metabolic changes, Dr. Marc Herlant of our staff administered 0.2 ml of lipo-adrenal cortex extract (Upjohn) to one group of 24-hour adrenalectomized rats while another group similarly treated was simultaneously submitted to stress of 6 hours duration. It was found that while the cortical extract alone caused essentially no thymus caryoclasia, the administration of the extract in conjunction with stress produced a marked caryoclastic reaction (Figs. 3 and 4). Thus from the preceding studies it is concluded that the metabolic effects of stress in the adrenalectomized animal are indeed capable of conditioning cortical hormones in so far as their thymolytic and gluconeogenesis actions are concerned.

Other investigations of peripheral conditioning have been performed by Engel and his co-workers (1949). Using the level of blood urea as an indicator of gluconeogenesis in the bilaterally nephrectomized and adrenalectomized rat they have demonstrated that the administration of glucose entirely inhibits the gluconeogenesis normally produced by adrenal cortex extract. This observation is in agreement with the previous report of Long *et al.* (1940). The results of these experiments have prompted the authors to conclude that "the increase in protein catabolism after adrenal cortex extract is readily modified by changing the internal metabolic environment."

Further evidence in favor of the conditional action of some corticoids lies in the sensitization to the toxic effects of desoxycorticosterone acetate (DCA) by an increased intake of sodium chloride (Selye, 1942; Selye and Stone, 1943; Selye and Hall, 1943; 1944). Under these experimental conditions DCA administered for a prolonged period has been shown to produce renal and cardiac enlargement, nephrosclerosis, myocarditis, hypertension, and periarteritis nodosa. That the conditioning effect of sodium chloride on the production of the above lesions is due to the sodium ion was indicated by the complete protection which the administration of acidifying salts such as ammonium chloride afforded (Selye, Hall, and Rowley, 1945) while potassium chloride was without protective activity (Masson and Beland, 1943; Selye, Mintzberg, and Rowley, 1945). Furthermore, the complete absence of sodium chloride from the diet has recently been shown to prevent the development of these manifestations of DCA treatment (Selye *et al.*, 1949). It has also been shown (Selye and Hall, 1944) that additional steroids such as progesterone and acetoxypregnenolone are similarly conditioned by the sodium chloride of the diet. Recently Dr. Selye has demonstrated (Selye, 1950) that a naturally occurring mineralocorticoid, 17-hydroxy-11-desoxycorticosterone (Reichstein's compound S) in daily doses of 2.5 mg in the presence of 1% sodium chloride in the drinking water and unilateral nephrectomy, can produce hypertension and pathological changes identical to those following chronic DCA treatment. It should be mentioned at this point that whereas the site of sodium chloride conditioning is peripheral, the hypertension and pathological changes produced in the intact rat by both prolonged stress and crude anterior pituitary preparations are also enhanced by an increased sodium chloride intake. The conditioning action of sodium chloride has recently been extended still further to include both ACTH and cortisone, which have been shown by Drs. Ducommon and Guillemin to produce hypertension in the rat in the presence of an excessive intake of sodium chloride.

From the facts outlined above it has been concluded that the salt-active corticoids probably produce organ changes by virtue of their sodium retaining property and that the conditioning of this action by sodium chloride is through the provision of adequate amounts of this ion.

Several recent publications lend support to the statement previously made that one corticoid can condition the peripheral actions of another. It was reported from this laboratory by Selye at the Adrenal Cortex Conference, 1949, that large continued doses of cortisone inhibit the development of arthritis and periarteritis produced by chronic DCA treatment. Similarly Perera and Pines (1949) have observed that adrenal cortical extract inhibits, at least to some extent, the elevation of blood pressure which follows the administration of DCA to hypertensive patients. Furthermore, the testoids through their renotropic action are known to protect the kidney from the nephrosclerosis induced by DCA as well as the damage caused by sublimate intoxication or ureteral ligation (Selye, 1940; Selye and Friedman, 1941; Selye and Rowley, 1944). The significance of this conditioning is emphasized by the fact that castration of the male rat, a procedure which of course removes the main source of androgen, produces some sensitization to the toxic effects of DCA (Selye and Pentz, 1943; Selye *et al.*, 1943; 1944).

Discussion and Conclusions

Although the reactions of an organism subjected to stress remain physiological under normal circumstances, the theory has been advanced that under certain conditions this process may become pathological and result in the development of hypertension as well as nephrosclerosis, arthritis, encephalitis, disseminated periarteritis and carditis of the rheumatoid type. If such is the pathogenesis of these diseases, it is obvious that the determination of the factors responsible for the development of an abnormal reaction to stress is of the utmost importance.

From the foregoing observations it seems justified to conclude that the state of the internal metabolic environment can condition or modify the function of the active pituitary-adrenal system and furthermore that this conditioning can occur at either the pituitary, adrenal, or peripheral level. Some of the factors creating a milieu which fosters the development of an abnormal defense reaction are now known. For example, protein seems to enhance the ability of the pituitary to produce corticotropin, hyperglycemia apparently increases the reactivity of the adrenal to some corticotropin preparations, while sodium chloride sensitizes the peripheral cells to the toxic effects of mineralocorticoids. On the other hand, not all the known conditioning factors potentiate the production of pathological lesions. Some indeed have prevented their development. Thus androgens, ammonium chloride, and high carbohydrate diets have been shown to protect against the development of nephrosclerosis and hypertension following LAP or DCA treatment.

Although we do not yet know how to produce or prevent selectively the development of any one adaptive disease in the experimental animal, this is the ultimate purpose of our research on factors conditioning the activity of the pituitary-adrenal system. Thus it is hoped that we may learn not only how the human body's reaction to nonspecific stress becomes perverted to produce selectively any one of the diseases of adaptation, but also how such an abnormal response may be prevented.

Acknowledgments

Some of the studies discussed in this review have been supported by the following organizations: American Heart Association, The Commonwealth Fund of New York, and the National Cancer Institute of Canada. The ACTH was kindly supplied by the Armour Company, Chicago.

The author wishes to express his appreciation to his colleagues Drs. Paris Constantinides and Claude Fortier whose advice and assistance have been invaluable in the preparation of this manuscript.

REFERENCES

Abelin, I., *Helv. Physiol. Pharmacol. Acta,* **3,** 71 (1945).

Benna, R. S., and Harvard, E., *Endocrinology,* **36,** 170 (1945).

Berman, D., Hay, E. C., and Selye, H., *Proc. Can. Physiol. Soc.,* **20** (1945).

Constantinides, P., *Federation Proc.,* **9,** 25 (1950).

Constantinides, P., Fortier, C., and Skelton, F. R., *Endocrinology,* in press.

Dontigny, P., Hay, E. C., Prado, J. L., and Selye, H., *Am. J. Med. Sci.,* **215,** 442 (1948).

Dugal, L. P., and Thérien, M., *Endocrinology,* **44,** 420 (1949).

Dumm, M. E., Ovando, P., Roth, P., and Ralli, E. P., *Proc. Soc. Exptl. Biol. Med.,* **71,** 468 (1949).

Engel, F. L., Schiller, S., and Pentz, E. I., *Endocrinology,* **44,** 458 (1949).

Fortier, C., Skelton, F. R., Constantinides, P., Timiras, P., Herlant, M., and Selye, H., *Endocrinology,* **46,** 21 (1950).

Fortier, C., Constantinides, P., and Skelton, F. R., *Proc. Can. Physiol. Soc.,* October, 1950.

Giroud, A., Santa, N., and Martinet, M., *Compt. rend. soc. biol.,* **134,** 23 (1940).

Giroud, A., Santa, N., Martinet, M., and Bellon, M. T., *Compt. rend. soc. biol.,* **134,** 100 (1940).

Giroud, A., and Martinet, M., *Compt. rend. soc. biol.,* **135,** 1344 (1941).

Giroud, A., Martinet, M., and Bellon, M. T., *Compt. rend. soc. biol.,* **135,** 514 (1941).

Hay, E. C., Prado, J. L., and Selye, H., *Can. J. Research,* **E 26,** 212 (1948).

Henriques, S. B., Henriques, O. B., and Selye, H., *Endocrinology,* **45,** 153 (1949).

Ingle, D. J., Ginther, G. B., and Nezamis, J., *Endocrinology,* **32,** 410 (1943).

Long, C. N. H., Katzin, B., and Fry, E. G., *Endocrinology,* **26,** 309 (1940).

Masson, G., and Beland, E., *Rev. can. biol.,* **2,** 487 (1943).

Moya, F., Prado, J. L., Rodriguez, R., Savard, K., and Selye, H., *Endocrinology,* **42,** 223 (1948).

Perera, G. A., and Pines, K. L., *Proc. Soc. Exptl. Biol. Med.,* **71,** 443 (1949).

SELYE, H., *Brit. J. Exptl. Path.*, **17,** 234 (1936).
SELYE, H., *Can. Med. Assoc. J.*, **42,** 173 (1940).
SELYE, H., *J. Pharmacol. Exptl. Therap.*, **68,** 454 (1940).
SELYE, H., *Can. Med. Assoc. J.*, **47,** 515 (1942).
SELYE, H., "Stress," Acta Inc., Montreal, 1950.
SELYE, H., *Brit. Med. J.*, **1,** 203 (1950).
SELYE, H., The Josiah Macy Jr. Foundation, Adrenal Cortex Conference, 1949.
SELYE, H., and DOSNE, C., *Proc. Soc. Exptl. Biol. Med.*, **48,** 532 (1941).
SELYE, H., and FRIEDMAN, S. M., *Endocrinology,* **29,** 80 (1941).
SELYE, H., and STONE, H., *Proc. Soc. Exptl. Biol. Med.*, **52,** 190 (1943).
SELYE, H., and HALL, C. E., *Arch. Path.*, **36,** 19 (1943).
SELYE, H., HALL, C. E., and ROWLEY, E. M., *Can. Med. Assoc. J.*, **49,** 88 (1943).
SELYE, H., and PENTZ, E. I., *Can. Med. Assoc. J.*, **49,** 264 (1943).
SELYE, H., and HALL, C. E., *Am. Heart J.*, **27,** 338 (1944).
SELYE, H., and ROWLEY, E. M., *J. Urol.*, **51,** 439 (1944).
SELYE, H., SYLVESTER, O., HALL, C. E., and LEBLOND, C. P., *J. Am. Med. Assoc.*, **124,** 201 (1944).
SELYE, H., HALL, O., and ROWLEY, E. M., *Lancet,* **248,** 301 (1945).
SELYE, H., MINTZBERG, J., and ROWLEY, E. M., *J. Pharmacol. Exptl. Therap.*, **85,** 42 (1945).
SELYE, H., STONE, H., TIMIRAS, P. S., and SCHAFFENBURG, C. A., *Am. Heart J.*, **37,** 1009 (1949).
SELYE, H., and STONE, H., American Lecture Series, (1950).
SKELTON, F. R., FORTIER, C., and SELYE, H., *Proc. Soc. Exptl. Biol. Med.*, **71,** 120 (1949).
SKELTON, F. R., *Proc. Can. Physiol. Soc.*, October, 1950.
STEEPLES, G. L., and JENSEN, H., *Am. J. Physiol.*, **157,** 418 (1949).

CONTROL OF REGENERATION OF THE ADRENAL CORTEX IN THE RAT

DWIGHT J. INGLE

Research Laboratories, The Upjohn Company, Kalamazoo, Michigan

The adrenal cortex has the capacity to regenerate new cells when parts of it are destroyed. This is a review of our studies on regeneration of the adrenal cortex of the rat.

HISTORY

My first studies of the adrenals were started in 1929 and had to do with the transplantation of these glands. I learned, as others had discovered earlier, that the success of isoplastic grafts was determined by the genetic relationship between the donor and the host and that autoplastic grafts grew best. I was not the first to observe that a transplanted gland fails to grow in the presence of an intact gland. A hormonal insufficiency in the host is required for the successful regeneration of grafts of endocrine tissues. This principle was discovered by Halstead (1909) in studies of the parathyroid grafts in the dog.

From 1934 to 1938, it was my good fortune to be associated with Dr. George M. Higgins at the Mayo Institute of Experimental Medicine. During those years, we studied the viability of transplanted adrenal tissue under a number of different conditions. It was my task to excise the adrenal and attach it to some other tissue, usually the surface of the ovary. After recovery of the graft some time later, it was placed in a fixative and George Higgins carried on from there with the microtome and microscope. We learned that in addition to direct experimentation, the papers by Drs. L. C. Wyman and Caroline Tum-Suden of the Boston University Medical School provided a fertile source of information on the behavior of the transplanted adrenal gland, and on more than one occasion observations that were new and exciting to us proved to have been fully anticipated by our friends in Boston.

Observations (Ingle and Higgins, 1938c) on the adrenal transplant which relate to regeneration can be summarized as follows. Degeneration began immediately in the center of the gland and extended peripherally, and in 48 hours the entire gland, except for the capsule and a narrow zone of glomerulosa, had been destroyed. During this time, fibrous and vascular connections were established between the host and the capsule of the graft. Within 72 hours the blood capillaries penetrated the capsule and a wide band of polymorphonuclear leucocytes, lymphocytes and red blood cells had entered the necrotic area of the graft. Regeneration of cortical tissue began at this time. Centers of proliferation of cortical cells appeared within the capsule and the zona glomerulosa, and gave rise to islands of cortical tissue which grew and coalesced so that 5 or 6 weeks after transplantation a mass of cortical tissue approximately equal to that of the original graft was restored. When one adrenal was transplanted in the presence of a contralateral intact adrenal, regeneration was suppressed although the graft remained viable. Regeneration did not occur in the absence of the anterior pituitary. The presence of the capsule was essential for regeneration of the graft. In fact, the transplantation of the capsule together with an adherent layer of glomerulosa cells was sufficient for regeneration of cortical tissue.

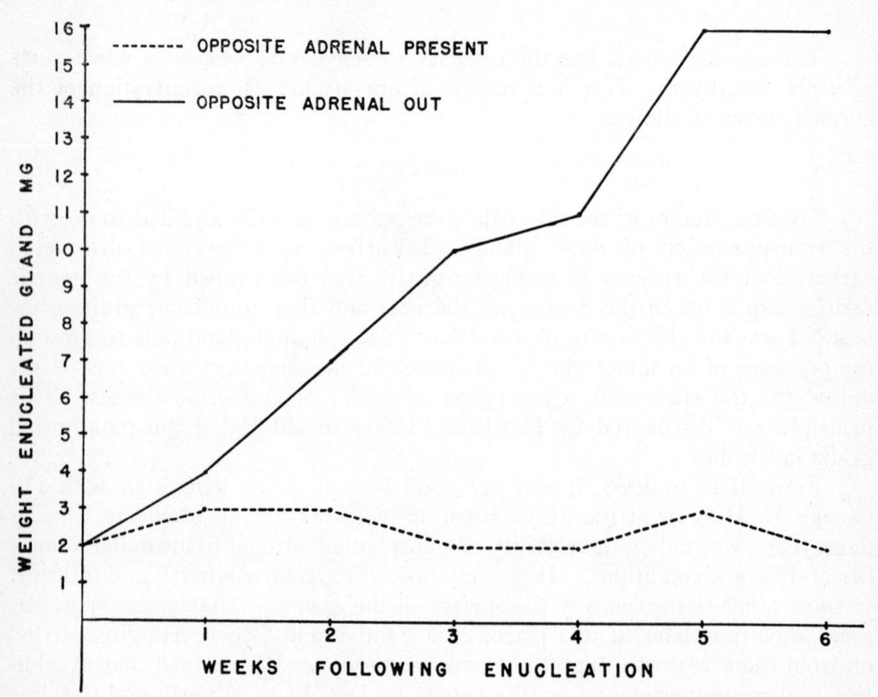

FIG. 1. The effect of the presence or absence of an intact adrenal upon the extent of regeneration of an enucleated adrenal. One rat is represented at each point. Pituitary present.

The description of adrenal demedullation in the rat by Evans (1936) led us to adopt this procedure for the study of adrenal cortex regeneration. The tip of the gland is clipped away and the gland is extruded from its capsule by a gentle squeezing action using fine, smooth, curved iris forceps. The capsule, with a layer of glomerulosa cells and occasionally a few cells from the fasciculata, remains attached to the pedicle of the gland. At the suggestion of the late Professor Edgar Allen, we adopted the term "enucleation" as descriptive of this procedure.

In our report (Ingle and Higgins, 1938a) on regeneration of the enucleated adrenal, it was shown that when both adrenals are enucleated there is rapid regeneration of cortical cells in the remnant. When one adrenal is enucleated and the other removed, cortical regeneration proceeds within the enucleated gland, and this single regenerating gland grows to a larger size than either of a pair of enucleated glands. When one adrenal is enucleated and the other left intact, there is little or no regeneration within the remnant. The effect of the presence or absence of an intact adrenal upon the extent of regeneration in the contralateral enucleated adrenal is shown in Fig. 1. The remnant remains viable and is stimulated to regenerate by the removal of the intact adrenal at any later time. The administration of large amounts of adrenal cortex extract suppresses the regeneration of an enucleated adrenal

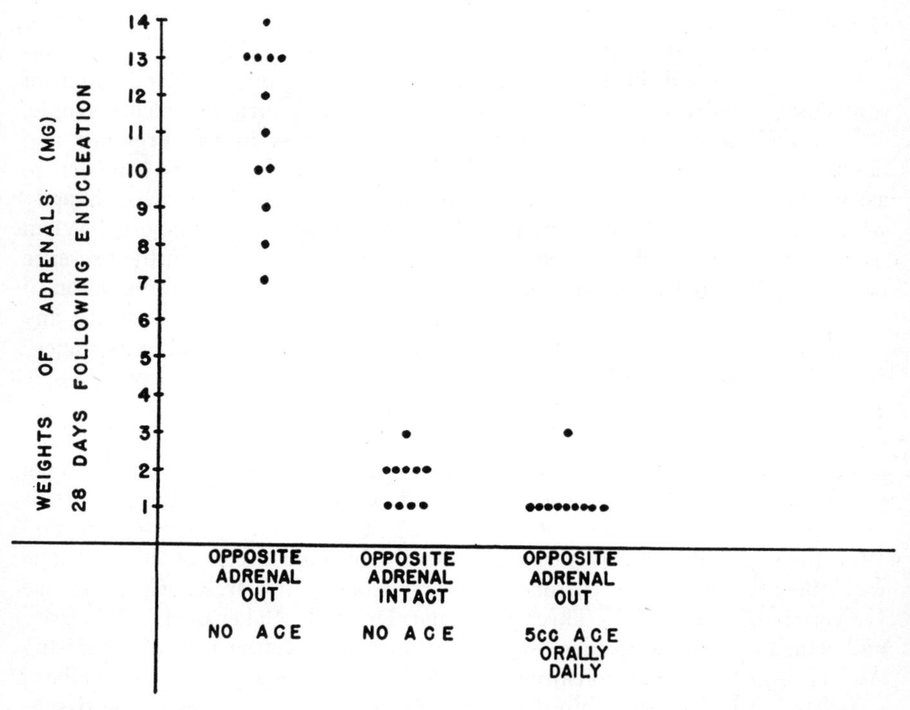

FIG. 2. Suppression of regeneration of the enucleated adrenal by the administration of large amounts of adrenal cortex extract. Pituitary present. Three weeks.

as effectively as does the presence of an intact adrenal. This effect of cortical hormone deficiency and excess upon regeneration is illustrated by the data on weights of enucleated adrenals which are summarized in Fig. 2. Finally, hypophysectomy completely suppressed regeneration of the enucleated adrenal. The parenchymal cells of the remnant became more atrophic following hypophysectomy. Regeneration of the enucleated adrenal occurs in partially hypophysectomized rats in which the infundibulum has been destroyed.

Elsewhere it was shown that stress (Higgins and Ingle, 1938; Ingle and Higgins, 1938b) increased the extent of regeneration in the enucleated adrenal when the opposite gland had been removed, but stress failed to stimulate regeneration in the presence of the opposite gland although this intact gland would increase greatly in size.

It was also shown (Ingle and Higgins, 1939) that the extent of regeneration was limited by the size of the remnant left at the time of enucleation. When the remnant was very small the regenerated gland, although representing hyperplasia of several hundred per cent, never regained a normal size.

All the foregoing observations are consistent with the following assumptions. (1) Regeneration of cortical tissue is controlled by the adrenocorticotropic hormone (ACTH) of the anterior pituitary. (2) Increased amounts of ACTH are secreted when the animal becomes adrenally insufficient. (3) The secretion of ACTH is insufficient to stimulate regeneration when the amounts of cortical hormone (endogenous or exogenous) are sufficient to meet the needs of the tissue. (4) The secretion of ACTH is further suppressed by cortical hormone excess. Our studies on the production of compensatory adrenal cortical atrophy during cortical hormone excess (Ingle, 1938a) and on the response of the cortex to stress in the presence and absence of the anterior pituitary (Ingle, 1938b) gave further support to assumptions (3) and (4). I am sure that we would have lived happily with this hypothesis if we had not tried to obtain more direct support. When we attempted to stimulate regeneration of the enucleated gland in the presence of an intact adrenal and in the hypophysectomized rat by the administration of the existing preparations of ACTH, we were unable to demonstrate any positive effects. Here was a simple question of fundamental importance: Is adrenal regeneration controlled by the single known adrenocorticotropic hormone of the anterior pituitary or are other factors essential?

At this point our studies on adrenal regeneration in the Mayo Institute came to an end, and the problem of the relationship of ACTH to adrenal regeneration was left dangling. Ten years passed before I returned to this problem. In the meantime ACTH was isolated in homogeneous form, and I became engaged in a series of studies of the biologic effects of this hormone with the brilliant young chemist who first isolated it, Dr. C. H. Li of the University of California. Three years ago, Dr. B. L. Baker of the University of Michigan joined us in these studies. After the exacting task of separating ACTH from the other constituents of California sheep pituitaries has been completed, it is dispersed into the body fluids of Kalamazoo rats whose tissues later continue the journey to Ann Arbor where Dr. Baker, a histochemist, has the final word as to what happened.

Current Studies

Our present studies of adrenal regeneration were planned to answer two questions: (1) In nonhypophysectomized rats, can regeneration of one enucleated adrenal be stimulated by Li's ACTH when the opposite adrenal is left intact? (2) In the hypophysectomized rat, can the enucleated adrenal be stimulated to regenerate by Li's ACTH? Our recent results support the answer "yes" to both questions and give rise to a third question which is now under investigation: (3) In the hypophysectomized rat, does the presence or absence of one intact adrenal modify the extent of regeneration in the opposite enucleated adrenal?

Nonhypophysectomized male rats of approximately 300 g weight were force-fed a medium carbohydrate diet. The animals were placed in activity-restriction cages and were given constant subcutaneous injections of either ACTH or saline by means of a continuous infusion machine which delivered 1 cc in 24 hours. The left adrenal was enucleated and the right adrenal was left intact in all the rats. It was found that the administration of 1 mg of ACTH daily had little or no effect upon the extent of regeneration of the enucleated gland although there was hyperplasia of the intact adrenal.

A dose of 3 mg of ACTH daily for 21 days stimulated regeneration of the enucleated gland to a significant extent. When the dose of ACTH was increased to 6 mg daily in 3 rats, dramatic changes occurred. The animals lost weight rapidly (up to 111 g in 21 days), developed glycosuria, one was dying and was posted on the nineteenth day of injection, and the remaining animals were extremely weak and incoordinate at the time they were killed on the twenty-first day.

At necropsy, the enucleated adrenals were regenerated and large. The enucleated adrenal of a control rat showed little regeneration as compared with the regenerated adrenal of a similar rat treated with 6 mg of ACTH daily.

The intact adrenals of these animals were the largest that we have ever seen, weighing 176, 215, and 286 mg. Although the outer portion of each of these glands consisted of viable cortical tissue, the inner portion was necrotic and hemorrhagic with masses of stagnant blood.

It was very interesting to find evidence that the adrenal cortices of the rat are capable of secreting such an excess of their hormones that the animal is literally brought to death by their action. The glycosuria, the rapid melting away of tissue in these force-fed rats, the fact that ACTH has no known metabolic effects in the absence of the adrenal glands (Ingle, Prestrud, and Li, 1948) all indicate a state of extreme hypercorticalism. To indulge in an anthropomorphism, these three rats might each have said "My adrenals kill me."

We have administered ACTH by constant subcutaneous injection to hypophysectomized rats having the left adrenal enucleated. Because of the slowed intestinal absorption that characterizes hypophysectomized rats, it was not possible to force-feed them. The injection of 3 mg of ACTH daily for 21 days caused a significant regeneration of the enucleated gland. When 6 mg per day were given, the weight loss of these animals was accelerated and it was necessary to kill the animals after 10 to 14 days. A significant amount of regenerated cortex, approximately as much as occurs in the presence of the pituitary, was found at necropsy.

A single ACTH factor is capable of stimulating regeneration of the enucleated adrenal of the hypophysectomized rat. We thought that this was true in the beginning, but our faith in the concept of a single ACTH factor was shaken by our failure to stimulate regeneration by divided injections (up to 5 daily) and because there were other attractive hypotheses postulating up to 4 principles affecting the adrenal cortex. Of course the finding that a single principle stimulates cortical regeneration does not rule out the possibility that other pituitary hormones may also affect the cortex. It does eliminate the argument that there *must* be something more than ACTH because ACTH failed to work.

New Problems

I stated earlier that our current studies were intended to answer three questions and that we have the answers to two of them. By this time a number of new problems are apparent. What is the mechanism whereby ACTH causes pathologic changes in the adrenal glands? Are these lesions related to the adrenal hemorrhage and necrosis occurring in various fulminating infections and other severe stresses in patients, to the adrenal changes seen in pantothenic acid-deficient rats and following other stresses in laboratory animals? Are all these changes due to excessive stimulation of the adrenal cortices during stress or are there other factors which act directly upon the adrenal?

What is the origin of new cells during regeneration? A number of investigators, including Dr. Higgins and myself, found evidence that the connective tissue of the capsule can be differentiated into secretory cells. This concept was so vigorously attacked by Greep and Deane (1949) that I came to feel that I was on the wrong side of the fence. Recently I have taken heart, for Dr. Baker, who looked at the role of the capsule in regeneration with skepticism, has excellent evidence that the primitive connective tissue cells of the fibrous capsule can give rise to parenchymal cells. The telling of this story will be left to Dr. Baker.

Since I have exhausted my information about adrenal cortex regeneration, it would be unfair to burden the reader with a complete list of questions which I cannot answer. Only one more question will be raised: It is now known that ACTH can stimulate the release of increased amounts of adrenal cortical hormone within a few seconds after it reaches its target organ. How does the effect of ACTH upon secretory activity relate to its effect upon the growth and morphology of the adrenal cortex?

Summary

Regeneration of the adrenal cortex is related to a physiological "need" for the cortical hormones and is apparently controlled specifically by the adrenocorticotropic hormone (ACTH) of the anterior pituitary. When one adrenal gland is enucleated and the contralateral gland removed, there is rapid regeneration from the remnant of the enucleated gland. Regeneration can be suppressed by the administration of large amounts of cortical extract or by leaving the contralateral gland intact. Regeneration of the cortical tissue does not occur in the hypophysectomized animal. The continuous injection of a homogeneous preparation of ACTH stimulated regeneration of the enucleated adrenal in hypophysectomized rats or in nonhypophysectomized rats having the contralateral gland intact.

REFERENCES

Evans, G., *Am. J. Physiol.,* **114,** 297 (1936).

Greep, R. O., and Deane, H. W., The adrenal cortex. *Ann. N. Y. Acad. Sci.,* **50,** 596 (1949).

Halstead, W. S., *J. Exptl. Med.,* **11,** 175 (1909).

Higgins, G. M., and Ingle, D. J., *Endocrinology,* **23,** 424 (1938).

Ingle, D. J., *Am. J. Physiol.,* **124,** 369 (1938a).

Ingle, D. J., *Am. J. Physiol.,* **124,** 627 (1938b).

Ingle, D. J., and Higgins, G. M., *Am. J. Med. Sci.,* **196,** 232 (1938a).

Ingle, D. J., and Higgins, G. M., *Endocrinology,* **23,** 419 (1938b).

Ingle, D. J., and Higgins, G. M., *Endocrinology,* **23,** 458 (1938c).

Ingle, D. J., and Higgins, G. M., *Endocrinology,* **24,** 379 (1939).

Ingle, D. J., Prestrud, M. C., and Li, C. H., *Endocrinology,* **43,** 202 (1948).

EXPERIMENTAL ADRENAL CORTICAL TUMORS*

GEORGE W. WOOLLEY

Roscoe B. Jackson Memorial Laboratory, Bar Harbor, Maine, and Sloan-Kettering Institute, Memorial Hospital Center, New York

Tumors of the adrenal cortex have been observed in man, and they and their secretions and related physical and psychological abnormalities have been noted for many years.

Tumors of the adrenal cortex in experimental animals have not until recently been observed with sufficient frequency to be studied in detail. I would like to call to your attention and to discuss adrenal cortical tumors in mice which can be predicted and can occur with high frequency.

In 1939, Dr. Arnold Speigel, in Germany, reported finding adrenal cortical tumors in three guinea pigs castrated and then kept to 2½—3 and 4 years of age before autopsy. No further reference to studies on adrenal cortical tumors in guinea pigs has been found. In this same year we reported the occurrence of adrenal cortical tumors and related mammary gland tumors in strain *dba* mice gonadectomized at birth and allowed to age without further experimental treatment (Woolley *et al.,* 1939). Further study showed that these tumors occurred in 100% of the animals, and they followed an interesting and simple procedure in their development (Fekete *et al.,* 1945; Woolley, 1949).

Nodular Hyperplasia

The earliest changes in the adrenal cortex of the female strain *dba* mouse, which is here used as the example, consisted of the appearance of groups of small, densely arranged cells immediately below the capsule. Eventually these cells spread between the cells of the zona glomerulosa, zona fasciculata, and cells soon to be described. These subcapsular cells were polyhedral in shape, having deeply staining round nuclei and slightly basophilic cytoplasm. As they extended between the cell columns of the zona fasciculata they became spindle shaped and strikingly resembled the cells of the ovarian stroma. As the subcapsular cells increased in number by mitosis, they were seen first to interrupt the normal arrangement of the cortical cells and after invading, to replace the cells of all three zones. Soon after the appearance of the subcapsular cells, cells which had undergone hypertrophy were found. These have been termed type A and type B cells, respectively (Woolley and Little, 1945a). After undergoing hypertrophy the cytoplasm

* This work has been aided by grants to the Roscoe B. Jackson Memorial Laboratory by the Commonwealth Fund, The Anna Fuller Fund, The Jane Coffin Childs Fund for Medical Research, The National Advisory Cancer Council, and the American Cancer Society through the Committee on Growth.

accumulated lipoid, stained lighter with hematoxylin and eosin, and the cell outlines became more distinct. The nuclei became large and vesicular. Such cells were occasionally found in the process of cell division. Later these cells formed nests which were surrounded by spindle-shaped cells. By increase in size and number of cell components, the first wedge-shaped areas changed into round nodules and often involved large parts of the cortex. In some cases the enlarging nodules projected toward the center of the gland and pushed the medulla into an eccentric position. In other cases, the nodules caused bulging of the surface. In a number of glands enlargements occurred in both directions. Still further growth resulted in the invasion and finally the breaking-through of the capsule, after which a mushroom-like growth occurred and resulted in the involvement of the surrounding adipose tissue. In extreme cases, large areas of surrounding tissues were involved. Since circulation was seriously disturbed, necroses, thromboses, and calcification in the central area have been seen to occur. The same adrenal frequently contained several abnormal nodules which showed different phases of the changes described. Even in the most extreme involvement, some normal cortical and and medullary tissue was always present. The advanced changes were visible in the gross, enlarging the gland, and giving it a nodular outline.

The same nodular hyperplasia occurred in gonadectomized male and female mice, although at any age period the changes were more extensive and more advanced in the females than in the males.

Adrenal Cortical Carcinoma

In extending our observations we observed the occurrence of adrenal cortical carcinomas in strain *ce* mice gonadectomized 1 to 3 days after birth (Woolley *et al.*, 1943). Further study showed that they developed in the following manner (Fekete and Little, 1945; Woolley and Little, 1945a,c,d,e).

The preliminary adrenal cortical changes toward these tumors were very similar to the early changes toward nodular hyperplasia and have been described above. The smaller carcinomas were dense cellular nodules composed of atypical cells showing frequent mitotic figures. They were imbedded in the outer region of the cortex, in areas already disorganized with changes toward nodular hyperplasia. Expansion into the medulla was evident as the nodules enlarged. Eventually the medulla and the remaining noncarcinomatous cortical tissue were spread out more or less as a sheet a few cells thick, first over much of the tumor and later, because of the large size of the tumor, over but part of the outer surface. The growths were fairly rounded in their early stages, but as they attained a size of 1.5 cm or more in diameter, protuberances often developed. The rounded contour was also modified by the adjacent kidney. Not only did the kidney modify the tumor, but the tumor often modified the shape of the kidney. Invasion of the kidney was not noted despite the close spatial relationship. Invasion of the liver and metastasis to the lung have been observed.

The tumors were made up predominantly of two types of tissue. Type I had large polyhedral cells with large oval nuclei containing chromatin granules of medium size. These cells were diffusely arranged. This tissue was not encapsulated, although the surrounding cell columns and sinusoids, which were compressed by the expanding growth, in some places gave this impression. Type II had small cells, cuboidal in shape and arranged in rows. The cells had large, dark staining, oval nuclei with evenly distributed coarse chromatin granules. A small amount of slightly basophilic cytoplasm surrounded each nucleus. The cells resembled the follicular cells of young ovarian follicles.

Adrenal cortical carcinomas first appeared in 100% of the ovariectomized females at 6 months of age and were present in 100% of the succeeding age groups. None was observed in the intact females of this strain examined. The same tumors occurred with a high frequency in gonadectomized males and not in intact males. As with nodular hyperplasia, the tumors of the male tended to develop later and in this case with a lower bilateral frequency, than in females.

These tumors have also been transplanted into males, females, gonadectomized males, and gonadectomized females. The tumors have specific sex hormone requirements in the host in some instances. One tumor grew successfully in only male mice for the first several transplant generations: eventually it grew in animals of either sex. Transplant tumors tend to maintain their individually different hormone-producing abilities; that is, some will produce androgenic-like hormones and others can be depended upon to produce estrogenic-like hormones through at least several transplant generations—all that have been tested (Woolley and Little, 1947; Woolley, 1949).

Primary tumors of the adrenal cortex have recently been observed in castrated rats. These tumors have not been described as malignant.

We and others have observed that early gonadectomy, and even gonadectomy itself, is not essential even though helpful for adrenal cortical tumor occurrence in experimental animals. Gardner (1941) observed adrenal cortical tumors in mice ovariectomized 40 to 60 days of age, and we observed nodular hyperplasia in mice ovariectomized at 9 and 15 months of age (Woolley *et al.,* 1941). Adrenal tumors have been infrequent in most strains of mice if the animals had not been gonadectomized. Thus, Slye observed 4 adrenal cortical tumors in 33,000 autopsies (Slye *et al.,* 1921). Some spontaneous and transplantable tumors have been observed in *Bagg albino* mice and Kirschbaum observed cortical tumors in normal strain *NH* mice, although tumor formation was enhanced by gonadectomy (Dalton *et al.,* 1943; Frantz *et al.,* 1947; Frantz and Kirschbaum, 1948; King *et al.,* 1949). This latter observation is a fascinating one since apparently the ovary aged prematurely in this strain of mice, forming a spontaneous partial castration.

Other Cortical Tumors

Cortical tumors of types other than the above have been observed in the mouse. One tumor which attains a size somewhat larger than that of the adrenal itself occurs in intact mice at advanced ages. It seems to be formed from the subcapsular type A cell, without the second stage, cell hypertrophy. We have observed these tumors in intact mice where this is the only primary tumor and also in adrenals where part of the adrenal is abnormal with nodular hyperplasia.

Atypical cortical growth of another type and one with some tumor properties has been observed in old intact mice. In this case, groups of cortical cells, particularly in the zona fasciculata and zona glomerulosa have come out from under the normal control of the body and have produced tumors which usually stay fairly well within the normal confines of the cortex. In still older age these cells often hyalinize rather than grow progressively. The reason for these changes is not known.

Cortical Tumor Relationships

It is known that the tumors of the adrenal cortex, nodular hyperplasia, and adrenal cortical carcinoma are active secreting tumors. In the sex hormone field there is evidence that they secrete androgenic substances, and estrogenic substances including progesterone-like hormones. The type of hormone produced varies with the strain of mouse under observation. It may be almost exclusively androgenic, almost exclusively estrogenic, or in certain cases both hormones may be produced by the same tumor. These tumors can cause full development of accessory reproductive organs and lead to the development of secondary tumors such as of the uterus and of the mammary glands.

Certain groups of gonadectomized mice exhibit tumors of the anterior lobe of the hypophysis regularly after the occurrence of the adrenal cortical tumors. The evidence is that these tumors are related to the basophilic cells and are active secreting tumors of the pituitary—possibly forming the animal counterpart of Cushing's syndrome in man (Dickie, 1948; Dickie and Woolley, 1949; Woolley and Little, 1945e; Woolley, 1948, 1949).

Efforts have been made to try to control and prevent the occurrence of the adrenal cortical tumors by various steroid and protein hormones. One pellet of diethylstilbestrol, 4 to 5 mg in weight, 25% diethylstilbestrol in cholesterol, implanted when the gonadectomized mice are about 2 months of age is sufficient to protect the animals against the adrenal cortical tumors to at least 14 months of age (Woolley and Little, 1946). King found that 0.5 gamma diethylstilbestrol per day in the diet was sufficient to give protection. Calorie-restricted mice, with modified, i.e., lowered mammary tumor incidence, still exhibited adrenal cortical tumor formation (King *et al.*, 1949). Strong androgens, such as testosterone propionate or testosterone dipropionate, or 17-methylandrostanediol, 3α 17α and pronounced estrogens such as alpha-estradiol, estriol, or estrone prevented the formation of these tumors. Desoxycorticosterone acetate, progesterone, *cis*-testosterone, and a number of other hormones failed to prevent tumor formation (Woolley and Chute, 1947).

King *et al.* failed to find the nodular hyperplasia of caloric restricted mice responsive to a gonadotropin (1947) or to corticotropin (1948), facts which led him to postulate that this tissue was either not responsive to these influences or less responsive than ovarian tissue. Our analysis of the effect of pituitary hormone fractions on adrenal cortical tumor formation is not far enough along for presentation at this time.

Conclusion

A working hypothesis to account in general for these tumors is possibly a simple one. There is known to be strengthening of pituitary hormones following gonadectomy. These hormones reacting on the adrenal cortex, a tissue arising from the genital ridge, might well be the inciting cause of the adrenal tumor formation. Adrenal cortical tumor secretions in steady excess might well be the inciting cause of tumor formation of the anterior lobe of the pituitary.

Finally, one need at the moment is further study of the adrenal tumors to determine whether or not these tumors produce excess cortical substances other than the sex hormones. Another need, of course, is experimental therapy to determine the best methods of controlling these cortical tumors once they are formed.

REFERENCES

Dalton, J. A., Edwards, J. E., and Andervont, H. B., A spontaneous transplantable, adrenal cortical tumor arising in a strain C mouse. *J. Natl. Cancer Inst.*, **4**, 329-338 (1943).

Dickie, Margaret M., Histological appearance of pituitary tumors in experimental F_1 reciprocal hybrid mice. *Acta union intern. contre le cancer*, **6**, 252-253 (1948).

Dickie, Margaret M., and Woolley, George W., Spontaneous basophilic tumors of the pituitary glands in gonadectomized mice. *Cancer Research*, **9**, 372-384 (1949).

Dorfman, R. I., and Gardner, W. U., Metabolism of steroid hormones. Excretion of estrogenic material by ovariectomized mice bearing adrenal tumors. *Endocrinology*, **34**, 421-435 (1944).

Fekete, E., Woolley, George W., and Little, C. C., Histological changes following ovariectomy in mice. I. *dba* high tumor strain. *J. Exptl. Med.*, **74**, 1-8 (1941).

Fekete, E., and Little, C. C., Histological study of adrenal cortical tumors of gonadectomized mice. *Cancer Research*, **5**, 220-226 (1945).

Frantz, M. J., Kirschbaum, A., and Casas, C., Endocrine interrelationships and spontaeous tumors of the adrenal cortex. *Proc. Soc. Exptl. Biol. Med.*, **66**, 645-646 (1947).

Frantz, M. J., and Kirschbaum, A., Androgenic secretion by tumors of the mouse adrenal cortex. *Proc. Soc. Exptl. Biol. Med.*, **69**, 357 (1948).

Gardner, W. U., Estrogenic effects of adrenal tumors of ovariectomized mice. *Cancer Research*, **1**, 632-637 (1941).

King, J. T., Casas, C., and Carr, C. J., Effect of corticotropin on ovariectomized C3H mice bearing adrenal adenomas. *Federation Proc.*, **7**, 64 (1948).

King, J. T., Visscher, M. B., and Casas, C., Effect of gonadotropin on the function of adrenal cortical tumors in ovariectomized restricted C3H mice. *Federation Proc.*, **6**, 142 (1947).

King, J. T., Casas, C. B., and Visscher, M. B., The influence of estrogen on cancer incidence and adrenal changes in ovariectomized mice on caloric restriction. *Cancer Research*, **9**, 436-437 (1949).

Kirschbaum, A., Frantz, M., and Williams, W. Lane, Neoplasms of the adrenal cortex in non castrate mice. *Cancer Research*, **6**, 707-711 (1946).

Slye, M., Holmes, H. F., and Wells, H. G., Primary spontaneous tumors in the kidneys and adrenals of mice. *J. Cancer Research*, **6**, 305-336 (1921).

SMITH, F. W., The relationship of the inherited hormonal influence to the production of adrenal cortical tumors by castration. *Cancer Research,* **8,** 641 (1948).

SMITH, F. W., Castration effects on the inherited hormonal influence. *Science,* **101,** 279 (1945).

SPEIGEL, ARNOLD, Über das Auftreten von Geschwülsten der nebennierenrinde mit vermännlichender Wirkung bei fruhkastrierten meerschweinchenmannchen. *Virchow's Arch. path. Anat.* 305, Band, 2, Heft. seite 367.

WOOLLEY, GEORGE W., FEKETE, E., and LITTLE, C. C., Mammary tumor development in mice ovariectomized at birth. *Proc. Natl. Acad. Sci.,* **25,** 277-279 (1939).

WOOLLEY, GEORGE W., FEKETE, E., and LITTLE, C. C., Differences between high and low breast tumor strains of mice when ovariectomized at birth. *Proc. Soc. Exptl. Biol. Med.,* **45,** 796-798 (1940).

WOOLLEY, GEORGE W., FEKETE, E., and LITTLE, C. C., Effect of castration in the dilute brown strain of mice. *Endocrinology,* **28,** 341a-343a (1941).

WOOLLEY, GEORGE W., FEKETE, E., and LITTLE, C. C., Gonadectomy and adrenal neoplasms. *Science,* **97,** 291 (1943).

WOOLLEY, GEORGE W., and LITTLE, C. C., The incidence of adrenal cortical carcinoma in gonadectomized female mice of the extreme dilution strain. I. Observations on the adrenal cortex. *Cancer Research,* **5,** 193-202 (1945a).

WOOLLEY, GEORGE W., and LITTLE, C. C., The incidence of adrenal cortical carcinoma in gonadectomized female mice of the extreme dilution strain. II. Observations on the accessory sex organs. *Cancer Research,* **5,** 203-210 (1945b).

WOOLLEY, GEORGE W., and LITTLE, C. C., The incidence of adrenal cortical carcinoma in gonadectomized male mice of the extreme dilution strain. *Cancer Research,* **5,** 211-219 (1945c).

WOOLLEY, GEORGE W., and LITTLE, C. C., The incidence of adrenal cortical carcinoma in gonadectomized female mice in the extreme dilution strain. III. Observations on the adrenal glands and accessory sex organs of mice 12 to 24 months of age. *Cancer Research,* **5,** 321-327 (1945d).

WOOLLEY, GEORGE W., and LITTLE, C. C., The incidence of adrenal cortical carcinoma in male mice of the extreme dilution strain over one year of age. *Cancer Research,* **5,** 506-509 (1945e).

WOOLLEY, GEORGE W., and LITTLE, C. C., Prevention of adrenal cortical carcinoma by diethylstilbestrol. *Proc. Natl. Acad. Sci.,* **32,** 239-240 (1946).

WOOLLEY, GEORGE W., and LITTLE, C. C., Transplantation of an adrenal cortical carcinoma. *Cancer Research,* **6,** 712-717 (1947).

WOOLLEY, GEORGE W., and CHUTE, ROSANNA, Effect of steroids on adrenal cortical tumor formation in mice. *Trans. 16th Meeting of Conference on Metabolic Aspects of Convalescence.* Josiah Macy Jr. Foundation, New York, 1947.

WOOLLEY, GEORGE W., and DICKIE, MARGARET M., Genetic and endocrine factors in adrenal cortical tumor formation. *Cancer Research,* **7,** 722 (1947).

WOOLLEY, GEORGE W., Physiological relationships of pituitary tumors in experimental F_1 reciprocal hybrid mice. *Acta de union intern. contre le cancer,* **6,** 265-267 (1948).

WOOLLEY, GEORGE W., The adrenal cortex and its tumors. *Ann. N. Y. Acad. Sci.,* **50,** 616-626 (1949).

STUDIES ON THE SITE AND MODE OF ACTION OF THE ADRENAL CORTEX IN PROTEIN METABOLISM*

FRANK L. ENGEL
Duke University, Durham, North Carolina

Since the demonstration by Long, Katzin, and Fry in 1940, that adrenal cortical extract increased the nitrogen excretion and liver glycogen of the fasted rat, much effort has been expended in an attempt to localize the site of action of the hormone of the adrenal cortex at some point or points along the metabolic pathway between protein and carbohydrate. The degradation of protein itself, the inhibition of protein synthesis, the deamination of amino acids, the conversion of three carbon precursors to glucose and glycogen, the conversion of glucose itself to glycogen and the reverse process, and extrahepatic carbohydrate utilization have all been considered in this regard (Long, 1942; White, 1948). Data have been presented at one time or another which may be considered consistent with an action at each one of these points, but none have been conclusive. The most widely accepted viewpoint at present is that the adrenal cortex does have a role in the mobilization and catabolism of endogenous protein as well as in the utilization of carbohydrate. The general acceptance of this interpretation, however, has led to the widespread belief that an increased rate of protein catabolism (or a decreased rate of protein anabolism) is a necessary consequence of adrenal cortical hormone action under all circumstances and that the changes in nitrogen metabolism after stress are examples of this direct consequence. The purpose of this report is to summarize our own investigations on the site and mode of action of the adrenal cortex in protein metabolism and to attempt to achieve therefrom a unified concept concerning the role of the adrenal in certain physiological processes.

Our approach to the problem has had the following objectives, each of which will be considered in this review: (1) the development of a method by which small changes in protein metabolism could be measured quantitatively and the time relations in hormone action established; (2) the more precise localization of the site of action of the adrenal cortex in protein metabolism both in terms of the level in protein metabolism at which it acts and in terms of the tissues predominantly involved; (3) the consideration of some of the factors which influence the magnitude and time of occurrence of the change in protein metabolism.

---

* These studies have been supported by grants from the Committee on Research in Endocrinology of the National Research Council, the American Cancer Society, administered by the Committee on Growth of the National Research Council, and the Duke Research Council.

Method of Measuring Small Changes In Protein Metabolism: Magnitude and Duration of Action of Small Doses of ACE

Most studies concerned with endocrine regulation of nitrogen metabolism have depended on the use of the conventional nitrogen balance technique or on measurements of tissue nitrogen (Long, Katzin and Fry, 1940; White, 1949). Although both of these methods have yielded valuable information, they have not been capable of detecting rapidly changing rates of protein metabolism. To be accurate, nitrogen excretion studies must be carried out over at least 12 hours, particularly in small animals. Under these circumstances the effects of a single small dose of hormone may be missed or modified by the development of compensatory changes due to antagonistic hormone action during the later hours. To overwhelm these compensatory responses larger amounts of hormone may have to be used, in which case overdosage effects may become significant. Furthermore, it has become increasingly evident, particularly with respect to the adrenal cortex, that hormone action is much more rapid than had been thought previously. Effects of adrenocorticotropic and cortical hormones on the eosinophil count (Thorn and Forsham, 1949), on the lymphocyte count (Dougherty and White, 1944), on liver fat (Levin, 1949), on ketonemia (Bennett *et al.*, 1948) and on uric acid excretion (Thorn and Forsham, 1949) have been demonstrated as early as 20 minutes to 5–6 hours after injection of hormone. The need for a simple method for measuring rapid changes in nitrogen metabolism for this problem is thus apparent.

We have made use of the measurement of the rate of urea formation in the nephrectomized rat for this purpose (Engel, Pentz, and Engel, 1948). The method depends on the assumptions that urea is an end product of protein catabolism and is freely diffusible throughout the body water. The rates of protein catabolism and anabolism, of deamination of amino acids and of urea synthesis are factors in determining the rate of accumulation of urea.

The details of the technique used in these studies have been described in an earlier publication (Engel, Pentz, and Engel, 1948). In most experiments the animals were bilaterally nephrectomized in the afternoon, fasted overnight, and measurements of the increment of the blood urea concentrations were made at 1- to 3-hour intervals. The rate of urea formation, expressed as milligrams of urea nitrogen per 100 g body weight per hour was calculated on an assumed mean body water of 63% of the body weight.

Using this technique it was found that the rates of accumulation of urea N overnight and during successive 3-hour periods thereafter in a large series of nephrectomized rats remained constant for at least 24 hours postoperatively. Under average conditions changes of the order of 0.75 mg of N per 100 g body weight per hour can be detected with accuracy by this method in a series of 9 to 10 animals. Russell and Cappiello (1949) have confirmed and improved on these results. They find that by taking samples immediately after nephrectomy they could detect the same differences as above in one-half or less the number of animals.

The effects of Upjohn's adrenal cortex extract (ACE) were then investigated. As can be seen in Fig. 1 the injection of ACE subcutaneously had no detectable effect on urea formation during the first 3 hours but during the following 3 hours a significant increase ranging from 0.6 to 1.0 mg N per 100 g body weight per hour took place. The ACE was given in divided doses at 0 and 1 hour beginning with 0.25 ml up to 1.0 ml each hour. With all doses above 0.25 ml a significant increment occurred during the second 3 hours, both as compared to the first 3 hours in each group as well as to the saline control. From these data it may be concluded that in the fasted nephrectomized rat an injection of adrenal cortical extract produces a transitory but highly significant increase in the rate of urea formation which becomes measurable about 3 hours after injection. Thus a change in nitrogen metabolism in response to cortical hormone treatment has been demonstrated to occur with the same rapidity as the other metabolic changes already noted.

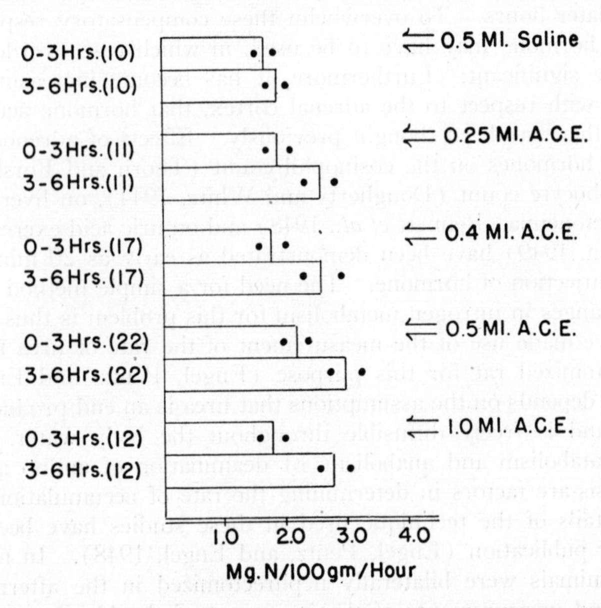

FIG. 1. Effects of injection of saline and ACE at 0 and 1 hour on urea formation in the nephrectomized rat.

LOCALIZATION OF THE SITE OF ACTION OF THE ADRENAL CORTEX IN PROTEIN METABOLISM

The increase in the rate of urea formation after ACE in the nephrectomized rat could result directly or indirectly from a change in the rate of one or more of the steps between protein and carbohydrate noted above.

It seemed reasonable to assume that if a changed rate of deamination of amino acids were a feature of adrenal cortical hormone action it would be detectable by an increased rate of urea formation following the injection of amino acids into the nephrectomized animal. That such does not seem to be the case is shown in Fig. 2. When a 10% mixture of the 10 essential amino acid plus glycine (Merck's VuJ, composition recorded in Engel, Schiller, and

Pentz, 1949) was injected intravenously in a dose of 9.4 mg of nitrogen per 100 g, a significant increase in urea formation occurred during the subsequent 3 hours. On the other hand, the same dose of amino acids administered to rats treated with adrenal cortical extract stimulated no greater rate of urea formation than when either one alone was given. The failure of the cortical hormone-treated rats to show an increased urea formation when amino acids were injected could be due to the fact that (1) the capacity of the liver for deamination was exceeded, (2) the maximal ability of ACE to promote deamination was already reached at the time of the amino acid injection, or (3) the injected amino acids suppressed an effect of ACE on stimulating

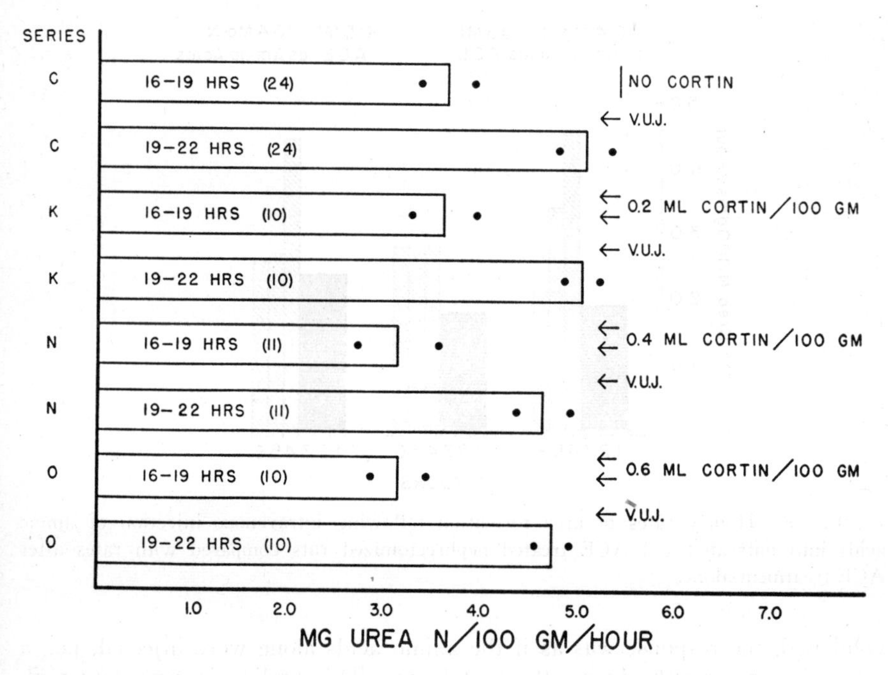

FIG. 2. Effects of ACE and intravenous amino acids (Merck's VuJ) on urea formation in the nephrectomized rat.

protein breakdown. The first possibility was readily ruled out by injecting twice the dose of amino acids. Under these circumstances the amount of urea formed during the following 3 hours was approximately doubled, indicating that the capacity of the liver for deamination was still considerable. The amount of urea appearing after the injection of 9.4 mg N as VuJ was quantitatively similar to that after ACE in these experiments and, as will be seen later in this review, considerably larger amounts of urea have been formed after ACE treatment under certain circumstances, indicating that explanation (2) is unlikely. The third possibility is supported by the following observation. In a previous study (Engel and Engel, 1946) it was found that the major part of the urea formed after an intravenous injection of amino acids appeared during the first hour, a finding confirmed by Russell and Cappiello (1949). It seemed possible that a separation of the amino acid and ACE effects on urea formation might be achieved by making hourly determinations during the period in which the amino acids or hormone were effective. The results are seen in Fig. 3. Injection of 10.4 mg of N as amino acids was followed by a prompt increase in urea which was most marked during the first hour after injection, less during the second hour, and no longer apparent during the third. One-half milliliter of ACE injected at the 0 and first hours resulted in a fairly steady increment in urea apparent during the 4th, 5th, and 6th hours respectively. When ACE and VuJ were

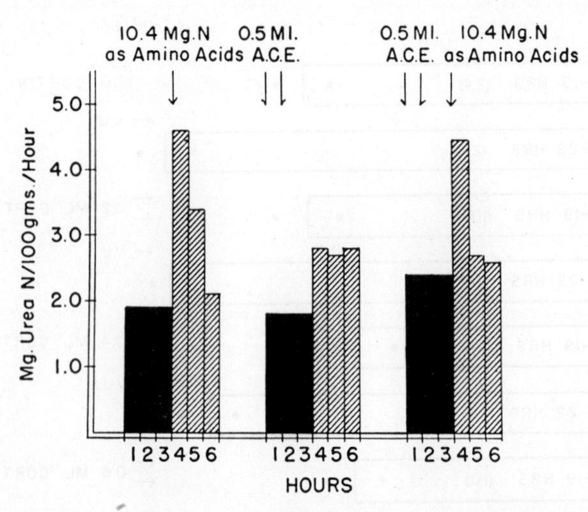

FIG. 3. Hourly rates of urea formation following intravenous injection of amino acids into untreated and ACE treated nephrectomized rats compared with rates after ACE treatment alone.

combined, the response was as if the amino acids alone were injected, i.e., a marked increase only during the first hour. This finding is consistent with the third possibility, but by itself it cannot be taken as more than suggestive evidence for this interpretation.

Further evidence against deamination being the central locus of action of ACE is found in the studies on adrenalectomized-nephrectomized rats given amino acids. Confirming an earlier observation of Evans (1941), it was found that the adrenalectomized-nephrectomized rat increased its rate of formation to the same degree as the nephrectomized rat given the same dose of amino acids (Bondy, Engel, and Farrar, 1949).

The apparent inhibition by amino acids of the stimulating effect of ACE on protein metabolism raised two fundamental questions concerning the mechanisms of this effect, neither of which can be answered with complete satisfaction. First, was the inhibitory effect of the amino acids related in any way to their ability to increase carbohydrate stores, this in turn having a secondary effect on protein metabolism? This seemed possible since the administration of amino acids to the ACE-treated rats was associated with a significant increase in liver glycogen compared to that in rats given ACE alone. Or was the apparent effect of the amino acids a direct one on protein, inhibiting protein breakdown or stimulating protein synthesis?

The glyconeogenetic potentialities of the VuJ amino acid mixture were found not to be a determining factor in the inhibitory action of amino acids. When the mixture was reconstituted to contain the same amino acids but in such proportions that 74% of the nitrogen was from nonglycogenic amino

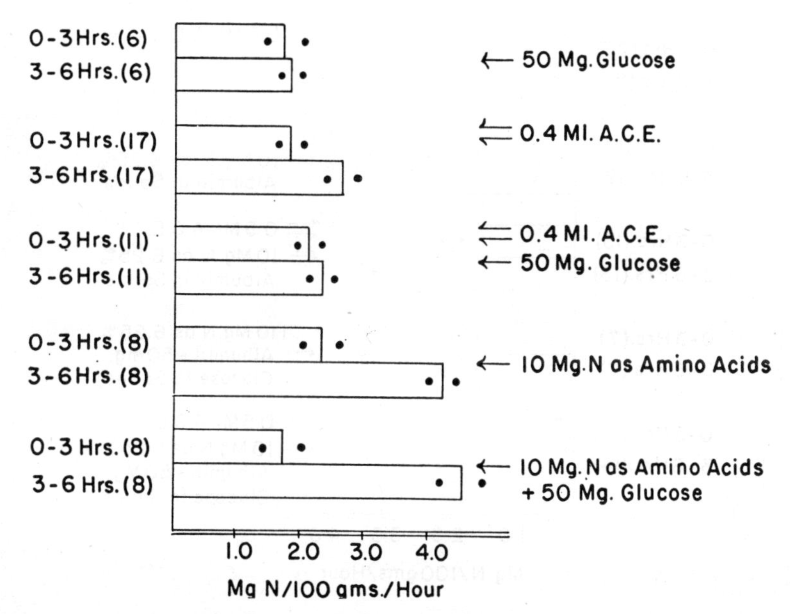

FIG. 4. Effects of intravenous glucose administration on urea formation after ACE and after amino acid injection in nephrectomized rats.

acids instead of the 70% glycogenic in VuJ, or when a single nonglycogenic amino acid, lysine, was used, no additive effect on urea formation from ACE and the amino acids was noted (Engel, Schiller, and Pentz, 1949). On the other hand, when glucose was given to the ACE-treated rats a complete inhibition of the ACE effect on urea formation also resulted (Fig. 4). The evidence that the inhibitory effect of glucose was at the level of protein rather than amino acids is good, and from these data one may speculate that the inhibitory effect of amino acids may also be exerted at this level. The evidence may be summarized as follows:

1. Glucose does not decrease the amount of urea formed after an injection of amino acids (Fig. 4). This is true even if much larger amounts of glucose (250 mg per 100 g) are given.

2. If 10 mg of nitrogen as human serum albumin are given to the nephrectomized rat, no increase in urea formation occurs during the subsequent 3

hours, unless ACE is given. In the latter case a considerable increase occurs, which is completely obliterated by giving glucose with the albumin (Fig. 5).

3. The nephrectomized rat maintained on a diet low in potassium (for reasons noted below) shows an increase in urea formation during the 3 hours after injection of 10 mg of N as rat plasma or human serum albumin even without ACE treatment. This is presumably related to some defect in incorporating protein into cells because of the potassium deficiency. In the presence of glucose no increase in urea formation occurs.

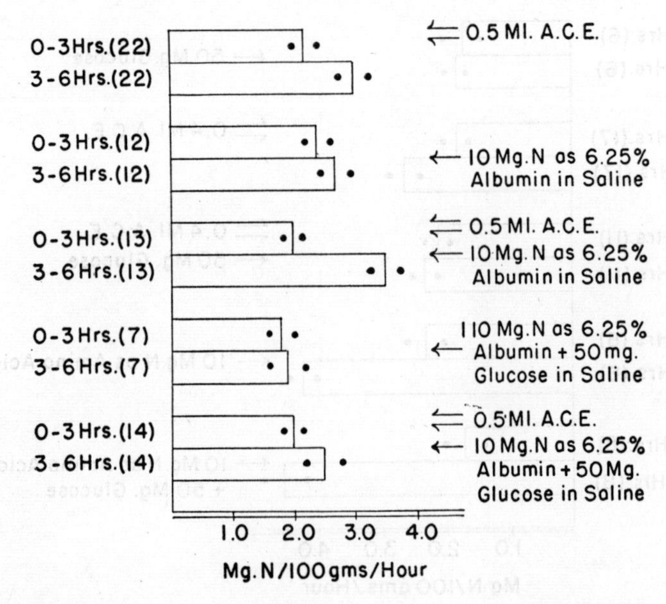

FIG. 5. Effects of intravenous albumin and glucose injection on urea formation in control and ACE-treated nephrectomized rats.

4. Glucose prevents the increased accumulation of amino acids in the plasma of the eviscerated rat treated with ACE (see below and Fig. 6). This last observation indicates that the liver is not necessary for the inhibitory action of glucose or, presumably, of amino acids on endogenous protein breakdown.

The fact that glucose inhibits the increase in urea formation after ACE and appears to be influencing protein metabolism directly focuses attention on protein as one site of action of the adrenal cortex. It is significant in this regard that glucose in the doses used here does not decrease the basal rate of urea formation but simply that increment associated with increased adrenal cortical activity. Further data tending to delimit the locus of action of the hormone to changes in protein degradation or synthesis may be found in the following experiments. As already stated and in contrast to what happens when amino acids are given, the injection of human albumin into the nephrectomized rat treated with ACE results in a significantly greater rate of urea formation than when either ACE or albumin are injected separately (Fig. 5). Without the use of tracers, of course, it is impossible to tell whether the urea which appeared came from the injected protein or elsewhere. In either case, though, it is likely that it came from protein since amino acid injection did not have this effect. When adrenalectomized rats were used (Bondy, Engel, and Farrar, 1949), equally suggestive results were obtained. In order to achieve a reasonable period of survival after combined adrenalectomy and nephrectomy it was found necessary to deplete the animals of potassium by feeding them a low potassium diet for at least 4 days and injecting desoxy-

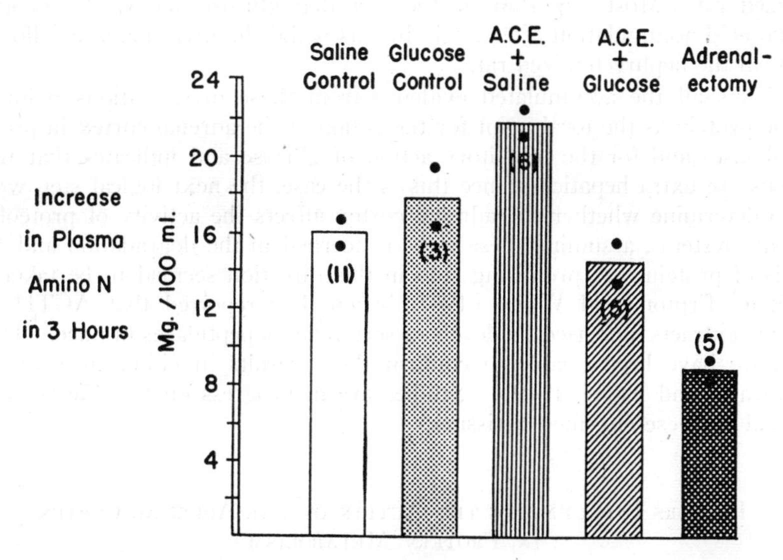

FIG. 6. Increase in plasma amino nitrogen levels of eviscerated rats following adrenalectomy or ACE treatment. Effect of glucose. (Data of P. K. Bondy.)

corticosterone acetate. Under these circumstances the adrenalectomized-nephrectomized rats actually survived longer than the nephrectomized animals, although when the former died they had lower blood urea levels (Bondy and Engel, 1947). Treatment with ACE decreased the survival time and increased the rate of urea accumulation. The significance of this observation will be considered later. When rat plasma was injected intravenously into the potassium-deficient nephrectomized rat, a significent increase in urea formation occurred in contrast to the results when either plasma or albumin were injected into nephrectomized rats previously on a stock diet. As noted above, this increase was abolished by glucose. In contrast, a similar injection of plasma into the adrenalectomized-nephrectomized rat was associated with no change in urea production unless ACE was given with the plasma. This confirms the long-held belief that the adrenalectomized animal has difficulty in using protein catabolically.

If protein breakdown or the ability to incorporate amino acids into protein were the locus of action of the adrenal cortex, one would anticipate that it would be possible to detect changes in the rate of amino acid accumulation in the plasma of the hepatectomized animal under the influence of the adrenal. This has been the case. Ingle, Prestrud, and Nezamis (1948) have shown that adrenalectomy lowered the rate of accumulation of amino acids in the eviscerated rat. Bondy (1949) has measured the effects of ACE and adrenalectomy on amino acid accumulation in this same preparation. As shown in Fig. 6 the eviscerate rat receiving ACE showed a significantly greater increase in plasma amino nitrogen in 3 hours than did its saline- or glucose-treated control, both of which in turn were greater than the increase in the adrenalectomized rat. Most important is the fact that glucose abolished this extra amino acid accumulation after ACE just as it did the urea increase following ACE in the nephrectomized rat.

Thus all the accumulated evidence from these investigations points to whole protein as the focal point for the action of the adrenal cortex in protein metabolism and for the inhibitory action of glucose and indicates that these actions are extra hepatic. Since this is the case, the next logical step would be to determine whether the adrenal cortex affects the activity of proteolytic enzyme systems, assuming these to be concerned in the degradation and synthesis of protein. A promising step in this direction seemed to be taken by Holman, Fruton, and White (1947) when they reported that ACTH and adrenal extracts increased the levels of serum aminopeptidases in mice. However, we have been unable to confirm these results in either mice or rats (Schwartz and Engel, 1949). Studies are in progress on the effects of the adrenal on these enzymes in tissues.

Factors Influencing the Action of the Adrenal Cortex in Protein Metabolism

The foregoing observations have demonstrated that the magnitude of the effect of ACE on protein metabolism is variable, at least in the nephrectomized or eviscerate rat. During the fasting state a measurable change in nitrogen metabolism occurs after ACE. In the protein-injected rat the increase in urea formation is even greater and in both cases the change is abolished if glucose is given. Amino acid mixtures, glycogenic or nonglycogenic, also inhibit the usual ACE response. While we have evidence that this inhibitory effect takes place at the level of whole protein and that in the case of glucose at least, the liver is not necessary, there is no ready explanation for the mechanism of the inhibition. It is not related simply to supplying an extra source of energy by the injection of the glucose or amino acids, for if an approximately isocaloric amount of a fat emulsion is injected intravenously, no inhibitory effect is seen (Engel, Schiller, and Pentz, 1949).

If supplying glucose inhibited the ACE effect, it seemed possible that contrariwise hypoglycemia might accentuate the effect (Engel, 1949). This proved to be the case, as seen in Fig. 7. Hypoglycemia-producing doses of insulin had no immediate effect on urea formation in the nephrectomized rat, as was also found by Russell and Cappiello (1949). On the other hand, a delayed increase did begin 3 hours after the insulin during the phase of spontaneous restitution of the blood sugar. This we believe was related to the stimulation of the pituitary-adrenal mechanism by insulin. When doses of 0.08 to 0.5 unit of insulin were given to rats previously treated with ACE, there was a very striking increase in urea formation, amounting to 3 to 4 times that seen after ACE alone. This change was most likely due to the hypoglycemia rather than the insulin *per se* because prevention of the hypo-

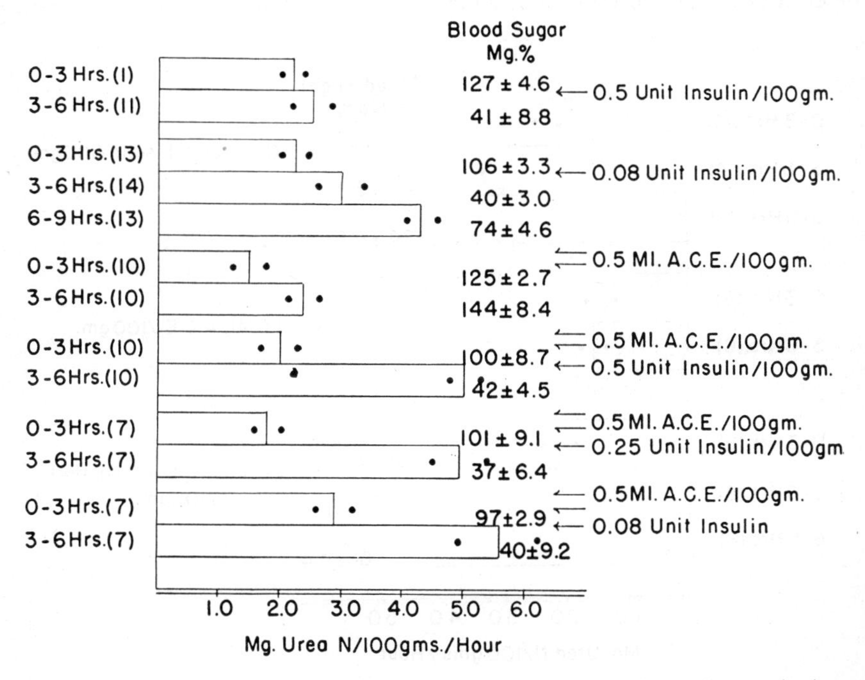

FIG. 7. Effects of insulin on the rates of urea formation in nephrectomized rats pretreated with ACE.

glycemia by glucose abolished the ACE effect. Not only did insulin hypoglycemia greatly increase the rate of urea formation after ACE, but it also hastened the occurrence of the ACE effect. It may be recalled, and some of the data are repeated in Fig. 8, that the ACE response in the fasted animal does not become measurable until 3 hours after injection of the hormone. Similarly, when insulin alone was given (Fig. 7), a delayed increase in urea formation occurred after 3 hours. In the previous experiments with insulin the ACE was given 3 hours before insulin. In the present experiment insulin and ACE were given stimultaneously and were followed by an immediate and marked increase in urea nitrogen which persisted through the following 3 hours. Thus, through the induction of hypoglycemia, the ACE effect on protein metabolism has been modified in two dimensions, time and magnitude.

Insulin, when given to the diabetic or in small doses to the normal animal, promotes protein synthesis. On the other hand, when hypoglycemia occurs, a stimulation of protein catabolism may be anticipated as part of the compensatory response to raise the blood sugar by increasing gluconeogenesis from protein. At the same time the adrenal cortex is stimulated, and it seems logical to conclude that the increase in protein catabolism is a direct consequence of the increased secretion of adrenal cortical hormone. However, this conclusion is not fully in accord with the facts just recorded. The urea increase after insulin and ACE occurs sooner and is vastly greater than that after the same or even twice the dose of ACE alone, so it seems unlikely that this response

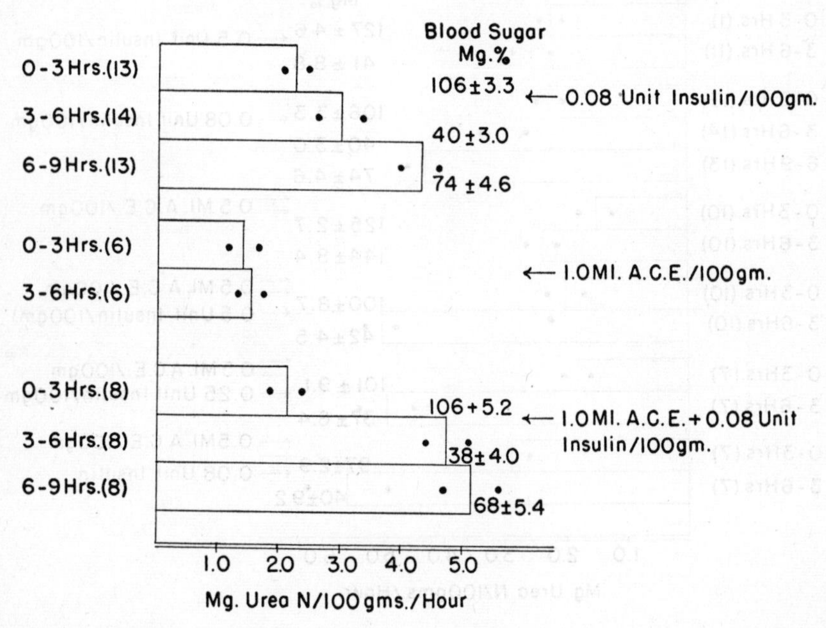

FIG. 8. Effects of simultaneous injection of insulin and ACE on urea formation in nephrectomized rats.

could be due to the adrenal alone (Fig. 1). This apparent discrepancy makes it necessary to introduce an additional factor into any consideration of the mechanism by which the adrenal cortex stimulates protein metabolism, namely, the poorly defined factor of "nonspecific stress" of which hypoglycemia might be considered as one example.

It is well known that the adrenal cortex hypertrophies (Tepperman, Engel, and Long, 1943) and becomes depleted of cholesterol and ascorbic acid (Long, 1947) when the organism is exposed to any stress. At the same time a negative nitrogen balance may develop if the stress is severe enough in the previously healthy organism but not in the adrenalectomized or hypophysectomized animal. An example of this failure of the adrenalectomized animal to change its rate of protein metabolism in response to stress is clearly shown in Fig. 9. This charts the rate of urea synthesis in adrenalectomized-nephrectomized and nephrectomized rats from the time of operation to death. The rate remains constant in the adrenalectomized animal but accelerates progressively in the animals with intact adrenals. That this latter response to stress in animals with intact adrenals is not simply a result of increased adrenal secretion was shown by Ingle, Ward, and Kuizenga (1947) who reported that adrenalectomized rats maintained on a constant dose of ACE did show the usual increase in nitrogen excretion after leg

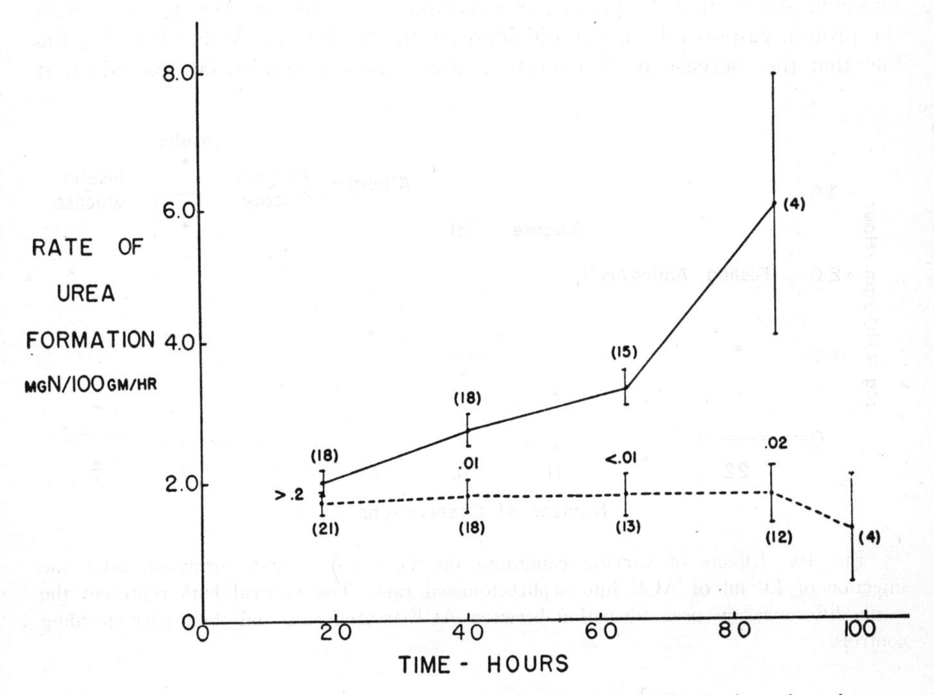

FIG. 9. Rate of urea accumulation in nephrectomized (solid line) and nephrectomized-adrenalectomized rats (broken line) from the time of operation to death. Rats maintained in low potassium diet for 4 days before nephrectomy. Reprinted from *Proc. Soc. Exptl. Biol. Med.*, **66,** 104 (1947).

fractures. From this and other studies Ingle (1949) has proposed the important and basic concept that the adrenal cortex is necessary but not responsible for the increased nitrogen excretion as well as certain other phenomena after stress. Related phenomena are the diabetogenic effects of stilbesterol (Ingle, 1944) and crude APE (Long, Katzin, and Fry, 1940) in adrenalectomized partially pancreatized rats maintained on subdiabetogenic doses of ACE, the glycogenic effect of anoxia in the adrenalectomized rat maintained on subglycogenic doses of ACE (Langley and Clarke, 1942), and the hyperglycemic effect of typhoid vaccine in the adrenalectomized-alloxan diabetic ACE maintained mouse (Tobian and Edwards, 1949). This in turn raises the question as to whether each of the circumstances here described in which a constant dose of ACE induced an increase in urea formation was not an example of a stress-induced acceleration of protein metabolism occurring in the presence of threshold amounts of cortical hormone. These various situations are summarized in Fig. 10 which diagrams the varying responses to identical doses of ACE under a variety of circumstances. One might hypothesize that the essential factor determining the magnitude of the protein response was not the adrenal cortex but the size of the "stress factor," with fasting being considered as the minimum and insulin hypoglycemia the maximal stress in these particular experiments. The inability to accentuate the protein response by a twofold increase in the dose of ACE (Fig. 1), the fact that the increase in N excretion after stress generally far exceeds that

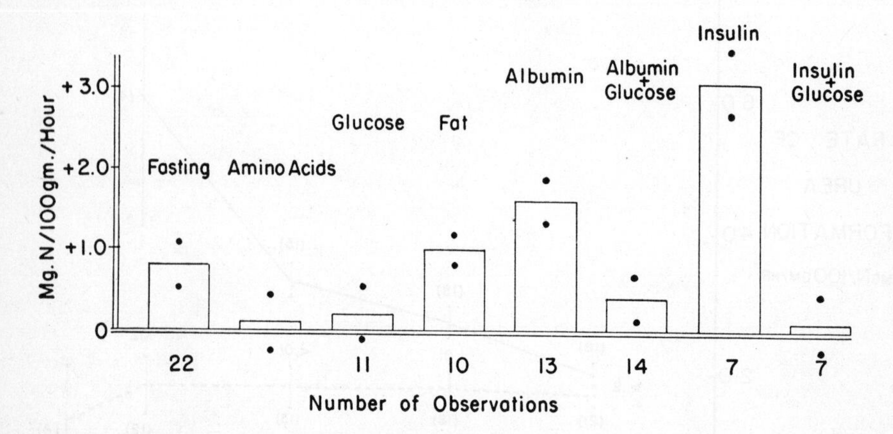

FIG. 10. Effects of varying conditions on the rates of urea formation after the injection of 1.0 ml of ACE into nephrectomized rats. The vertical bars represent the mean differences in urea formation between ACE-treated rats and their corresponding controls.

achieved even by overdosage with ACTH or cortical hormone (Noble and Toby, 1948), and the achievement of the full stress effect on N excretion in the traumatized adrenalectomized animal on constant ACE (Ingle, Ward, Kuizenga, 1947) supports such an interpretion. On the other hand, the fact that overdosage with cortical hormones can increase nitrogen excretion even in the absence of stress (Ingle *et al.,* 1946) and the following preliminary observation suggests that increasing amounts of available hormone are equally important in determining the magnitude of the response. In this experiment a stress was used (0.5 ml of formaldehyde subcutaneously) which was subthreshold in the sense that it did not accelerate the rate of urea production of the nephrectomized rat in 6 hours. If 1.0 ml of ACE was given with the stress, an increase in nitrogen metabolism took place which was just significantly greater than that from ACE alone during the first 3 hours and markedly so during the second 3 hours. Incomplete studies indicate that if twice the dose of formaldehyde is given, an increase in urea formation occurs without injecting ACE. It is not known yet whether ACE will increase this response or glucose inhibit it. In other words, with a constant "subthreshold" stress, adding ACE to that already circulating or stimulated to secretion brings out an acceleration of protein metabolism. Further studies are indicated, particularly in adrenalectomized animals receiving varying doses of hormone and subjected to different degrees of stress to establish these relationships more definitely.

It would seem possible, then, that the changing rate of protein metabolism which is commonly associated with increased adrenal cortical activity is in fact the resultant of two forces, one the adrenal cortex and the other as yet poorly defined but related in some general way to "stress." Either one, if great enough, may increase protein catabolism in the absence of the other. Thus large enough doses of cortical hormone will increase nitrogen excretion in the absence of stress (Ingle *et al.*, 1946) and conversely severe stress, such as fatal hemorrhage (Engel, Winton, and Long, 1943), or leg trauma (Noble and Toby, 1948), or even prolonged fasting (Ingle and Oberle, 1946), may increase protein metabolism in the adrenalectomized animal under the proper experimental conditions. Under most circumstances, however, neither one is effective without the other. In a certain number of cases, now, the increase in protein catabolism and/or decrease in protein anabolism which is the resultant of these two forces has been inhibited by glucose or amino acids. It is of interest in this regard that one other consequence of both stress and adrenal cortical activity, the development of a fatty liver, has also been found to be inhibited if glucose is given to the animal that is subjected to the stress or receiving ACTH (Levin, 1949). Furthermore, there is increasing evidence that the negative nitrogen balance which follows stress may be overcome by a large enough carbohydrate intake (Lathe and Peters, 1949; Werner, 1949).

The metabolic response to stress shows a typical pattern. It is characterized by an increased excretion of nitrogen, potassium, and phosphorus, a decreased tolerance to carbohydrate, a decreased sensitivity to insulin, and an increased rate of fat catabolism with an accumulation of fat in the liver and a tendency to ketosis. These changes are identical with those characterizing the state known as "starvation diabetes" and are similar to those following overdosage with cortical hormone with the exception that an increase in liver glycogen appears to be a feature of cortical hormone overdosage, but not of "stress." (The increase in liver glycogen of the hypoxic rat (Langley and Clarke, 1942) is an exception to this latter statement). None of these changes takes place in the adrenalectomized organism exposed to stress, but on the other hand there is evidence, reviewed above and by Ingle (1949), that they do take place in the adrenalectomized animal under the influence of maintenance doses of hormone. Thus the summating effects of stress and of adrenal cortical hormone activity probably have broader significance than would be anticipated from a consideration of protein metabolism alone. Nothing is known concerning the precise meaning and value of this metabolic response to the stressed organism nor what relation it bears to various mechanisms of resistance.

A more complete understanding of what constitutes the reaction to stress and injury and wherein the adrenal cortex is necessary for its full development is thus a focal point in our understanding of the physiological role of the adrenal cortex in health and disease. A breakdown of some of the adaptative mechanisms would appear to be a feature of a number of disease states. The effectiveness of large amounts of adrenal hormone in these conditions may be a manifestation of the summation of the two factors described above where the stress response by itself for some reason had been inadequate when taken with the normal maximal endogenous adrenal secretion. The adrenal cortex when fully active presumably makes possible the fullest development of all those processes concerned with maintaining the state of normalcy, whether it be normal growth (Ingle and Prestrud, 1949) or the complex set of metabolic and other events constituting the reaction to injury.

Summary

Experimental studies on the role of the adrenal cortex in protein metabolism have been reviewed and data presented which indicate that:

1. Changes in nitrogen metabolism following a single small injection of adrenal cortex extract into the fasted nephrectomized rat occur within 3 hours.

2. The site of action of the adrenal cortex in protein metabolism is at the level of whole protein rather than at the level of deamination of amino acids.

3. The so-called protein catabolic action of adrenal cortical extract can be inhibited by glucose or amino acids.

4. The magnitude of the so-called protein catabolic response to an injection of adrenal cortical extract may be increased and its time of occurrence accelerated by nonspecific stress.

It is suggested that the adrenal cortex and nonspecific stress play complementary roles in changing the rate of protein catabolism and that under most conditions maximal changes in protein catabolism do not take place unless both factors are at play.

Acknowledgment

The work reported here has been done in collaboration with Drs. P. K. Bondy, S. Schiller, and T. B. Schwartz, Mrs. Rosalie Green, and Mildred Engel, and Miss E. I. Pentz.

REFERENCES

Bennett, L. L., Kreiss, R. E., Li, C. H., and Evans, H. M., Production of ketosis by the growth and adrenocorticotropic hormones. *Am. J. Physiol.*, **152,** 210 (1948).

Bondy, P. K., unpublished observations, 1949.

Bondy, P. K., and Engel, F. L., Prolonged survival of adrenalectomized-nephrectomized rats on a low potassium diet. *Proc. Soc. Exptl. Biol. Med.*, **66,** 104 (1947).

Bondy, P. K., Engel, F. L., and Farrar, B., The metabolism of amino acids and protein in the adrenalectomized-nephrectomized rat. *Endocrinology,* **44,** 476 (1949).

DOUGHERTY, T. F., and WHITE, A., Influence of hormones on lymphoid tissue structure and function. The role of the pituitary adrenotrophic hormone in the regulation of the lymphocytes and other cellular elements of the blood. *Endocrinology,* **35,** 1 (1944).

ENGEL, F. L., Studies on the nature of the protein catabolic response to adrenal cortical extract. Accentuation by insulin hypoglycemia. *Endocrinology,* **45,** 170 (1949).

ENGEL, F. L., WINTON, M. G., and LONG, C. N. H., Biochemical studies on shock. 1. The metabolism of amino acids and carbohydrate during hemorrhagic shock in the rat. *J. Exptl. Med.,* **77,** 397 (1943).

ENGEL, F. L., and ENGEL, M. G., Urea synthesis from amino acids during hemorrhagic shock in the nephrectomized rat. *Am. J. Physiol.,* **147,** 165 (1946).

ENGEL, F. L., PENTZ, E. I., and ENGEL, M. G., On the use of the nephrectomized rat for the study of rapid changes in nitrogen metabolism. *J. Biol. Chem.,* **174,** 99 (1948).

ENGEL, F. L., SCHILLER, S., and PENTZ, E. I., Studies on the nature of the protein catabolic response to adrenal cortical extract. *Endocrinology,* **44,** 458 (1949).

EVANS, G., Deamination of DL-alanine in the adrenalectomized rat. *Endocrinology,* **29,** 737 (1941).

HOLMAN, H. R., WHITE, A., and FRUTON, J. S., Relation of adrenal cortex to serum peptidase activity. *Proc. Soc. Exptl. Biol. Med.,* **65,** 196 (1947).

INGLE, D. J., The diabetogenic effect of diethystilbestrol in adrenalectomized-hypophysectomized-partially depancreatized rats. *Endocrinology,* **34,** 361 (1944).

INGLE, D. J., Some studies on the role of the adrenal cortex in organic metabolism. *Ann. N. Y. Acad. Sci.,* **50,** 576 (1949).

INGLE, D. J., and OBERLE, E. A., The effect of adrenalectomy in rats on urinary non-protein nitrogen during forced-feeding and during fasting. *Am. J. Physiol.,* **147,** 222 (1946).

INGLE, D. J., SHEPPARD, R., OBERLE, E. A., and KUIZENGA, M. H., A comparison of the acute effects of corticosterone and 17-hydroxycorticosterone on body weight and the urinary excretion of sodium, chloride, potassium, nitrogen and glucose in the normal rat. *Endocrinology,* **39,** 52 (1946).

INGLE, D. J., WARD, E. O., and KUIZENGA, M. H., The relationship of the adrenal glands to changes in urinary non-protein nitrogen following multiple fractures in the force-fed rat. *Am. J. Physiol.,* **149,** 510 (1947).

INGLE, D. J., PRESTRUD, M. C., and NEZAMIS, J. E., Effect of adrenalectomy upon level of blood amino acids in the eviscerated rat. *Proc. Soc. Exptl. Biol. Med.,* **67,** 321 (1948).

INGLE, D. J., and PRESTRUD, M. C., The effect of adrenal cortex extract upon urinary non-protein nitrogen and changes in weight in young adrenalectomized rats. *Endocrinology,* **45,** 143 (1949).

LANGLEY, L. L., and CLARKE, R. W., The relation of the adrenal cortex to low atmospheric pressures. *Yale J. Biol. Med.,* **14,** 529 (1942).

LATHE, G. H., and PETERS, R. A., Protein sparing effect of carbohydrate in normal and burned rats. *Quart. J. Exptl. Physiol.,* **85,** 157 (1949).

LEVIN, L., Physical stress and liver fat content in the fasted mouse. *Federation Proc.,* **8,** 218 (1949).

LONG, C. N. H., A discussion of the mechanism of action of adrenal cortical hormones on carbohydrate and protein metabolism. *Endocrinology,* **30,** 870 (1942).

LONG, C. N. H., "Relation of Cholesterol and Ascorbic Acid to the Secretion of the Adrenal Cortex," in *Recent Progress in Hormone Research,* Vol. I, p. 99. Academic Press, New York, 1947.

LONG, C. N. H., KATZIN, B., and FRY, E. G., The adrenal cortex and carbohydrate metabolism. *Endocrinology,* **26,** 309 (1940).

NOBLE, R. L., and TOBY, C. G., The role of the adrenal glands in protein catabolism following trauma in the rat. *J. Endocrinol.*, **5**, 303 (1948).

RUSSELL, J. A., and CAPPIELLO, M., The effects of pituitary growth hormone on the metabolism of administered amino acids in nephrectomized rats. *Endocrinology*, **44**, 333 (1949).

SCHWARTZ, T. B., and ENGEL, F. L., The adrenal cortex and serum peptidase activity. *J. Biol. Chem.*, **180**, 1047 (1949).

TEPPERMAN, J., ENGEL, F. L., and LONG, C. N. H., A review of adrenal cortical hypertrophy. *Endocrinology*, **32**, 373 (1943).

THORN, G. W., and FORSHAM, P., "Metabolic Changes in Man Following Adrenal and pituitary Hormone Administration," in *Recent Progress in Hormone Research*, Vol. IV, p. 229. Academic Press, New York, 1949.

TOBIAN, L., JR., and EDWARDS, W. L. J., Exacerbation of alloxan diabetes in mice by injection of typhoid vaccine: Role of the adrenal gland. *J. Lab. Clin. Med.*, **34**, 487 (1949).

WERNER, S. C., A comparison in the same individual of the nitrogen loss following surgery with that produced before operation by an identical feeding program. *J. Clin. Invest.*, **28**, 818 (1949).

WHITE, A., "The Relation of Hormones to Protein Metabolism," in SAHYUN, M., *Proteins and Amino Acids in Nutrition*, pp. 236-265. Reinhold Publishing Corp., New York, 1948.

WHITE, A., "Integration of the Effects of Adrenal Cortical, Thyroid and Growth Hormones in Fasting Metabolism," in *Recent Progress in Hormone Research*, Vol. IV, p. 153. Academic Press, New York, 1949.

THE PROTECTIVE ROLE OF ADRENAL CORTICAL SECRETION IN THE HYPERSENSITIVE STATE*

THOMAS F. DOUGHERTY

University of Utah College of Medicine, Salt Lake City, Utah

Selye (1945, 1949) has emphasized the role of hypersecretion of the adrenal cortex in the development of diseases of connective tissues ("diseases of adaptation"). On the other hand, Rich and Gregory (1943, 1944, 1945, 1946) have obtained evidence which suggests that hypersensitivity phenomena play an important role in the development of connective tissue diseases. It is well known that the adrenal cortex responds with hypersecretion to a great variety of nonspecific stresses (Selye, 1937; Sayers *et al.*, 1944; Dougherty and White, 1947). Presumably, allergens may be included among the stresses which stimulate the adrenal cortex to hyperactivity. It is the purpose of this study to evaluate the relative importance of the two factors, adrenal cortical hypersecretion and hypersensitivity, in the manifestations of allergic phenomena.

Two experimental approaches have been employed: first, a quantitative analysis of the capacity of various adrenal cortical hormones to provide resistance to fatal anaphylactic shock and, secondly, histologic and cytologic studies of the alterations in connective tissue of hypersensitive animals subjected to allergic stimuli in the presence of varied amounts of adrenal cortical hormones.

Although it has been shown that adrenalectomized animals were more sensitive to anaphylactic shock than intact animals (Wyman, 1929; Wolfram and Zwemer, 1935; Weiser *et al.*, 1941), it was not possible to provide strong quantitative evidence that administration of either ACTH or adrenal cortical extracts increases resistance to the allergen dose (Leger *et al.*, 1948). It was felt that such quantitative data concerning protection against anaphylaxis might be obtained in mice since, although the intact sensitized mouse is the most resistant of animals (Weiser *et al.*, 1941; Mayer and Brousseau, 1946), the adrenalectomized sensitized mouse is highly reactive to antigen (Leger *et al.*, 1948). The therapeutic action of cortical hormone has been assayed on the basis of its ability to restore the resistance of adrenalectomized sensitized mice to that found in intact sensitized animals.

Experiments

All experiments which are to be described were planned in advance with the assistance of a mathematician adviser, and all experimental data were treated by the probit analysis method (Finney, 1947).

* These investigations were supported by the Life Insurance Research Foundation.

TABLE 1

Determination of Incidence of Anaphylactic Death in Sensitized Mice

Groups and Treatments	Dosage of Various Substances	Number of Animals	LD_{50} cc Horse Serum per 20-g Mouse	95% Fiducial Limits cc Horse Serum per 20-g		Index of Protection*	
				Upper	Lower		
I. Adrenex—untreated		323	0.00052	0.0006	0.00044	....	
II. Adrenex+doca	1 mg/20 g	69	0.00049	0.0007	0.00036	0.99	
III. Adrenex+lipo-adrenal extract	0.2 cc/20 g	164	0.0045†	0.025‡	0.0007‡	8.6	
IV. Adrenex+pyribenzamine (Deaths within 3 hr.)	2.5 mg/kg	217	0.102	0.165	0.0628	196	3.6
V. Adrenex+pyribenzamine (Total deaths within 18 hr.)	2.5 mg/kg	217	0.0285	0.0483	0.0168	55	
VI. Adrenex+cortisone acetate (10 min. before shock)	1 mg/20 g	40	0.0065	0.033	0.0014	12.5	12.9
VII. Adrenex+cortisone acetate (2 hr. before shock)	1 mg/20 g	26	0.084	0.139	0.051	162	
VIII. Adrenex+cortisone acetate (2 doses 2 hr. apart)	2 mg/20 g	45	>1.0	....	....	>2000	
IX. Intact-controls (sham operated)		82	>1.0	....	....	>2000	

* Treated adrenex as compared to non-treated adrenex.

† Midpoint range, LD_{50}.

‡ Upper and lower limits range of LD_{50}.

In several series of preliminary experiments an attempt was made to determine the degree of refractoriness of the CBA strain of mice to horse serum anaphylaxis (Dougherty, 1949). Mice were sensitized with 0.8 cc of horse serum in two divided doses and 21 days later were given shock doses of horse serum intravenously. The shock doses were varied from 0.01 to 1.0 cc per 20-g mouse. It was found that the strain of mice employed was completely insensitive to any dose from this wide dosage range (Table 1); not even a shock dose of 1.0 cc horse serum per 20-g mouse was capable of eliciting characteristic anaphylactic symptoms. The LD_{50} of horse serum in sensitized normal mice of strain CBA was thus established to be markedly over 1.0 cc; 1.0 cc horse serum did not even represent an LD_2, since no single lethal event was observed in 82 experiments.

In other experiments, the effect of adrenal cortical hyposecretion on the responsiveness of sensitized mice to anaphylactic shock doses was then investigated. Mice of the same strain were sensitized in the same manner as those of the preceding series. Shock doses were also administered 21 days after sensitization. Bilateral adrenalectomy was performed 2 hours before administration of the shock dose. The adrenalectomized horse serum-sensitized mice responded promptly with acute anaphylactic death to horse serum shock doses over a wide dosage range. The LD_{50} in adrenalectomized mice was found to be 0.00052 cc per 20-g mouse (95% fiducial limits; 0.0006 and 0.00044) (Table 1). Thus, absence of the adrenals was found to increase the anaphylactic hypersensitivity of the mouse by far more than 2000 times.

In order to ascertain whether the effect of adrenalectomy is due to the absence of adrenal cortical secretion, the cross experiment was undertaken to study the protective effectiveness of administration of adrenal cortical hormone (ACH) upon sensitized mice rendered susceptible to anaphylaxis by adrenalectomy. Sensitized adrenalectomized animals were given ACH in the form of lipo-adrenal (Upjohn) in a dose containing, according to bioassay, 0.5 mg of compound E (cortisone). The ACH dose was administered 2 hours before injection of the shock dose of horse serum, i.e., immediately after adrenalectomy. Under these experimental conditions, adrenal cortical hormone was found to exert a marked protective action. The LD_{50} of horse serum was established to be 0.0045 (range of LD_{50} of different groups of animals—0.025 and 0.0007) (Table 1). Hence, 0.5 mg of cortisone in form of the adrenal extract employed gives an approximately 8.6-fold protection against anaphylactic death.

The intact mouse has been reported to be insensitive not only to an antigen to which it has been sensitized (Wyman, 1929; Mayer and Brousseau, 1946), but also to histamine (Mayer and Brousseau, 1946). Histamine is considered to be one of a number of chemical mediators responsible for the clinical manifestations of the hypersensitive state; since it has pharmacological actions upon vascular muscle and capillary permeability it may be particularly involved in the mediation of pathological sequels of hypersensitivity.

For these reasons, it seemed appropriate to study the share of histamine in the anaphylactic response of the adrenalectomized mice by measuring a possible protective action of antihistaminic agents. Pyribenzamine (Ciba) was chosen as an antihistamine for these experiments and was administered 15 minutes before the shock dose of horse serum in a dose of 2.5 mg/kg to adrenalectomized sensitized mice pretreated as in the preceding experiments. This amount of pyribenzamine was found to increase the LD_{50} of horse serum to 0.102 cc (95% fiducial limits; 0.165 and 0.0628), i.e., it granted protection from anaphylactic shock to a greater degree than the lipo-adrenal extract, namely, 196-fold (Table 1).

However, pyribenzamine prevented only the acute anaphylactic death; the protection was effective for a period of 3 hours only. In the period between 3 and 12 hours the animals began to die in typical anaphylactic shock. If the delayed deaths are taken into account, the protective index was only 55 as referred to non-treated adrenalectomized sensitized mice (Table 1). The LD_{50} for the total acute and delayed deaths was 0.0285 cc of horse serum (95% fiducial limits; 0.0483 and 0.0168).

In order to determine whether the delayed shock manifestations after an initially protective pyribenzamine dose are due to the same mechanism as the early manifestations occurring in the absence of an antihistamine, ACH was administered toward the end of the initial period of pyribenzamine protection. Sensitized mice, shocked with horse serum 2 hours after adrenalectomy and 15 minutes after the standard pyribenzamine dose, received the same dose of lipo-adrenal as used in the previous experiments, within 3 hours after pyribenzamine administration. It was found that this procedure prevented the late minfestations and reduced the total mortality to the same level as in the 3-hour period for animals protected by pyribenzamine alone or in the 12-hour period for animals protected by an early ACH-dose alone.

The adrenal cortical extract employed contains a multiplicity of corticosteroids. The study of isolated pure corticosteroids was next undertaken. At the present time, experiments devoted to an analysis of the conditions of timing and dosage of cortisone administered intraperitoneally in the form of a crystalline suspension in saline have been completed. In the first series, 1 mg cortisone acetate administered 2 hours before the shock dose of horse serum was found to increase the LD_{50} to 0.084 (95% fiducial limits; 0.139 and 0.051) (Table 1). This signifies a 162-fold protection.

When the time interval is shortened and 1 mg cortisone is given 10 minutes prior to the shock dose, a lesser degree of protection is obtained (protective index: 12.5). When, however, in the third series a priming dose of 1 mg was given and a second 1-mg dose of cortisone was given 2 hours later, a high degree of protection was effected (LD_{50} over 1 cc horse serum) (Table 1).

Desoxycorticosterone acetate (Ciba) (1 mg/20-g mouse) was administered 2 hours before the shock dose of horse serum was given. It was evident (Table 1) that this hormone provided no degree of protection whatsoever.

Tissue Alerations of Mice Subjected to Anaphylactic Shock

Shortly after shock doses of horse serum in sensitized animals, extensive changes in the cardiovascular system occur. These changes, which were found within 15 to 45 minutes after horse serum, may be summarized as follows: (1) In the smaller vessels, particularly in the heart, lungs, kidney, and stomach wall, there was an extensive degeneration of the endothelial cells. This change consisted primarily of a swelling of the cytoplasm and formation of vacuoles in the nuclei. In many areas, particularly in the arterioles of the heart, small thrombi were observed. (2) The larger vessels exhibited splitting of the elastic membrane and edema of the media. Smooth muscle degeneration, primarily due to swelling of the individual cells and necrosis of the nuclei, was found to be a typical alteration in non-surviving animals. Finally, an extensive edema of the adventitia was observed in some of the larger vessels and in the connective tissue around the bronchioles.

These changes were found in varying degree in all groups of experimental animals. Of the changes in the vascular system, only the alterations in the endothelium were observed consistently in the intact shocked animals. It may be recalled that none of these animals died in acute anaphylaxis.

Administration of pyribenzamine prevented these alterations in the animals which survived the shock dose. However, in the group which died a delayed anaphylactic death, they were present in the same degree as in the adrenalectomized but non-pyribenzamine-treated animals. In cortisone-protected adrenalectomized animals, as in the intact animals, only the endothelial changes were observed. However, in animals which did not survive the shock dose, the vascular changes were as marked as in the adrenalectomized non-protected group. The most extreme changes were observed in the adrenalectomized animals given desoxycorticosterone acetate prior to the shock dose.

The alterations observed in the *respiratory system* were bronchiolar constriction, edema of bronchiolar cells, desquamation of the lining epithelial cells, emphysema and reticulo-endothelial reaction. All these alterations were found in all the insufficiently protected adrenalectomized animals which had been given shock doses of horse serum.

The principal changes in the *renal tissue* occurred in the capillary loops of the glomeruli; they consisted of swelling and vacuolization of the cytoplasm of the endothelium, thickening of the basement membrane in some of the animals, and glomerular hyperemia, which was consistently found in all the shocked animals but was most pronounced in the adrenalectomized non-protected groups.

As might be expected, the changes in the *lymphatic tissues* which are known to be produced by adrenal cortical secretion were present in the intact shocked animals but were completely absent in the adrenalectomized animals subjected to shock doses. Extensive lymphatic tissue alterations were observed in the cortisone-treated animals. There seemed to be no difference in the degree of lymphatic tissue reaction between those animals which were protected and those which succumbed to the shock dose.

Alterations in the Loose Connective Tissue

The loose or areolar connective tissue of the body is the prototype of all the more highly organized connective tissues. It was chosen for detailed study since it may be obtained readily, can be subjected to detailed cytological observation, and the fundamental changes occurring in loose connective tissue are similar to those found in other types of connective tissue (Dougherty, 1944). Investigations of this tissue revealed that acute inflammation took place following intravenous administration of horse serum to intact animals. Sublethal dosages of horse serum (0.00025 cc) given to adrenalectomized mice produced much more extensive cellular inflammatory responses than those observed in intact animals. The cellular inflammation was not as great, but was not absent, in the cortisone-protected adrenalectomized animals. Highly specific alterations in the fibroblasts occurred which consisted in the accumulation of numerous large red bodies in their cytoplasm.

In addition to the cellular manifestations of acute inflammation, changes in the amorphous ground substance took place which appeared to be related to the capacity of the animal to resist anaphylactic death.

Intact or cortisone-protected adrenalectomized animals exhibit accumulation of eosinophils in the loose connective tissue. This suggests the possibility that these cells may leave the blood and accumulate in the connective tissue under the influence of adrenal cortical hormones.

Discussion and Summary

The selection of the mouse as a test species for these investigations proved to be particularly fortunate for a demonstration of the interrelation between the adrenal cortical endocrine function and allergic manifestations and sequels of the hypersensitive state. In the work here presented, it turned out that the mouse, a species which is virtually incapable of responding with anaphylactic symptoms at any phase of sensitization and to any challenging dose of antigen, is as sensitive to anaphylaxis as is any other species, if it has been deprived of its adrenals. Thus, this work demonstrates that the adrenal cortex plays a dominant role in staving off the clinical phenomena of the anaphylactic reaction.

The diametrical contrast between the highly refractory intact and the highly sensitive adrenalectomized mouse invited quantitative study. It was found that the LD_{50} of the challenging antigen can be determined within a small limit of standard error. By use of this yardstick, it was possible to demonstrate that anaphylaxis is not an all-or-none phenomenon, but a graded response which, under otherwise uniform conditions of sensitization and choice of antigen, depends upon the challenging antigen dose and upon the quantitative relation between the doses of antigen and of protective agent, namely adrenal cortical hormone.

Although other signs of allergy such as the arthus phenomenon also were inhibited by administration of cortisone, such responses are not as readily subjected to quantitative study as anaphylactic death.

However, the study of these quantitative relations would be extremely inadequate if the measurement of the acute lethal effects were not supplemented by an investigation of the less manifest non-lethal sequels of the hypersensitivity reaction. This is important since the latter conditions are responsible for the cardiovascular and collagen tissue pathology of the hypersensitive state. The study of the tissues of hypersensitive mice challenged under the varied conditions of antigen dosage, endogenous or exogenous adrenal cortical hormone supply and other protective measures revealed that the allergic reaction invariably elicits numerous and characteristic tissue alterations. These alterations can be found to some degree in all sensitized and re-injected animals, whether intact or adrenalectomized, whether protected from or succumbing to the challenging antigen dose. The differences between the lesions in animals from groups with varying amounts of protection from lethal sequels are largely differences in degree.

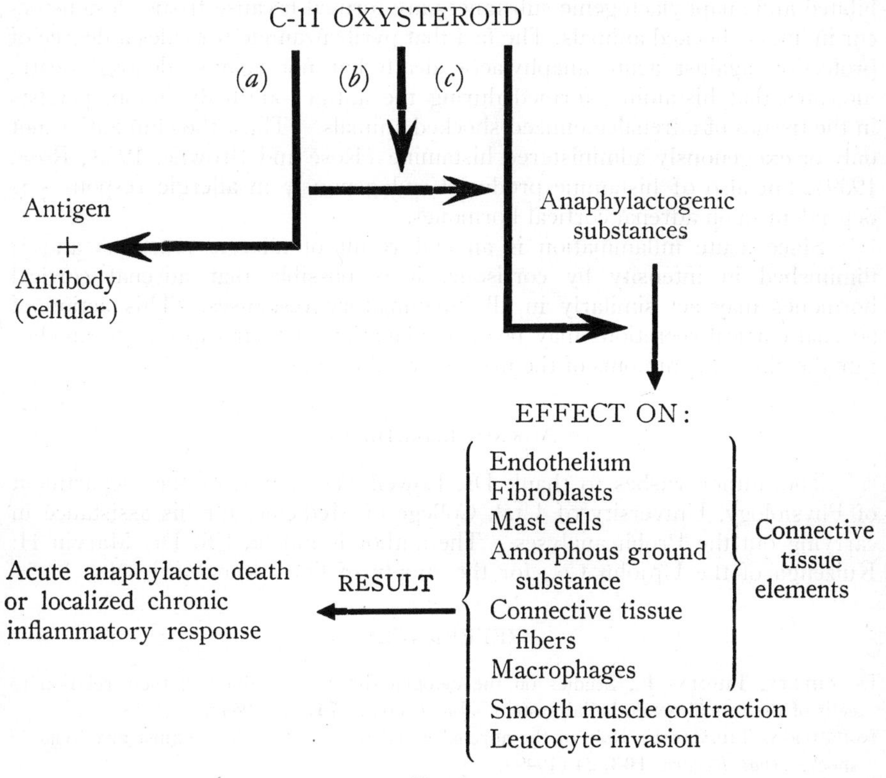

FIG. 1

Thus, it is clear from an analysis of both the studies of the protection against anaphylactic death and of the histological changes, that adrenal cortical hormones act in the connective tissues in a quantitative manner.

The direct mechanism of action of the C-11 oxysteroids in hypersensitive reactions has not as yet been ascertained. However, an analysis of the data obtained in the present experiments indicates that adrenal cortical hormones modify the extent of hypersensitive reactions by preventing the action of anaphylactogenic substances on cells (Fig. 1). Figure 1 depicts the steps in the hypersensitivity reactions and lists the tissues and cells affected by the anaphylactogenic substances. The term "anaphylactogenic substance" is used since several different agents such as histamine, heparin, hyaluronidase and necrosin (Mayer and Kull, 1947) could be released from cells following antigen-antibody union.

Adrenal cortical hormones could modify the anaphylactic response by acting in one or more of the following ways: by inhibiting union of antigen and antibody (Fig. 1*a*), diminishing the extent of this union (Fig. 1*b*) or decreasing production of or preventing the action of the anaphylactogenic substances (Fig. 1*c*).

It is clearly evident that the antigen-antibody union is not completely inhibited and anaphylactogenic substances are formed because tissues lesions occur in intact shocked animals. The fact that pyribenzamine provides a degree of protection against acute anaphylactic death but not against delayed death, indicates that histamine, formed during the antigen-antibody union, persists in the tissues of adrenalectomized shocked animals. Thus, the elimination not only of exogenously administered histamine (Rose and Browne, 1938; Rose, 1939), but also of histamine produced endogenously in allergic responses is dependent upon adrenal cortical hormones.

Since acute inflammation is an end result of allergic reactions and is diminished in intensity by cortisone, it is possible that adrenal cortical hormones may act similarly in all inflammatory responses. This action of adrenal cortical secretions may be an explanation of their capacity to ameliorate the clinical symptoms of the mesenchymal diseases.

Acknowledgments

The author wishes to thank Dr. Lowell Woodbury, of the Department of Physiology, University of Utah College of Medicine, for his assistance in carrying out the Probit analyses. The author is indebted to Dr. Marvin H. Kuizenga of the Upjohn Co., for the supply of lipo-adrenal extract.

REFERENCES

Dougherty, Thomas F., Studies on the cytogenesis of microglia and their relation to cells of the reticulo-endothelial system. *Am. J. Anat.*, **74**, 61 (1944).

Dougherty, Thomas F., Role of the adrenal gland in the protection against anaphylactic shock. *Anat. Record,* **103**, 24 (1949).

Dougherty, Thomas F., and White, Abraham, An evaluation of alterations produced in lymphoid tissue by pituitary-adrenal cortical secretion. *J. Lab. Clin. Med.,* **32**, 584 (1947).

Finney, D. J., *Probit Analysis.* Cambridge University Press, Bentley House, London, 1947.

Leger, Jacques, Leith, W., and Rose, Bram, Effect of ACTH on anaphylaxis in the guinea pig. *Proc. Soc. Exptl. Biol. Med.,* **69**, 465 (1948).

Mayer, R. L., and Brousseau, D., Antihistaminic substances in histamine poisoning and anaphylaxis of mice. *Proc. Soc. Exptl. Biol. Med.,* **63**, 187 (1946).

Mayer, R. L., and Kull, F. C., Influence of pyribenzamine and antistine upon the action of hyaluronidase. *Proc. Soc. Exptl. Biol. Med.,* **66**, 392 (1947).

RICH, ARNOLD R., Hypersensitivity to iodine as a cause of periarteritis nodosa. *Bull. Johns Hopkins Hosp.*, **LXXVII,** 43 (1945).

RICH, ARNOLD R., Hypersensitivity in disease—with especial reference to periarteritis nodosa, rheumatic fever, disseminated lupus erythematosus and rheumatoid arthritis. *Harvey Lectures, 1946-47,* **Ser. XLII,** 106 (1946).

RICH, ARNOLD R., and GREGORY, J. E., The experimental demonstration that periarteritis nodosa is a manifestation of hypersensitivity. *Bull. Johns Hopkins Hosp.,* **LXXII,** 65-88 (1943).

RICH, ARNOLD R., and GREGORY, J. E., Experimental evidence that lesions with the basic characteristics of rheumatic carditis can result from anaphylactic hypersensitivity. *Bull. Johns Hopkins Hosp.,* **LXIII,** 239-264 (1943).

RICH, ARNOLD R., and GREGORY, J. E., Further experimental cardiac lesions of the rheumatic type produced by anaphylactic hypersensitivity. *Bull. Johns Hopkins Hosp.,* **LXXV,** 115-134 (1944).

ROSE, BRAM, The effect of cortin and desoxycorticosterone acetate on the ability of the adrenalectomized rat to inactivate histamine. *Am. J. Physiol.,* **127,** 780 (1939).

ROSE, BRAM, and BROWNE, J. S. L., The distribution and rate of disappearance of intravenously injected histamine in the rat. *Am. J. Physiol.,* **124,** 412 (1938).

SAYERS, G., SAYERS, M. A., FRY, E. G., WHITE, A., and LONG, C. N. H., The effect of the adrenotrophic hormone of the anterior pituitary on the cholesterol content of the adrenals, with a review of the literature of adrenal cholesterol. *Yale J. Biol. Med.,* **16,** 361 (1944).

SELYE, HANS, Studies on adaptation. *Endocrinology,* **21,** 169 (1937).

SELYE, HANS, The general adaptation syndrome and the diseases of adaptation. *J. Clin. Endocrinol.,* **VI,** 117 (1945).

SELYE, HANS, Participation of adrenal cortex in pathogenesis of arthritis. *Brit. Med. J.,* 1129-1136, November, 1949.

WEISER, R. S., GOLUB, ORVILLE, and HAMRE, D. M., Studies on anaphylaxis in the mouse. *J. Infectious Diseases,* **68,** 97 (1941).

WOLFRAM, J., and ZWEMER, R. L., Cortin protection against anaphylactic shock in guinea pigs. *J. Exptl. Med.,* **61,** 9 (1935).

WYMAN, L. C., Studies on suprarenal insufficiency. VI. Anaphylaxis in suprarenalectomized rats. *Am. J. Physiol.,* **89,** 356 (1929).

MODIFICATION OF BODY STRUCTURE BY ADRENOCORTICAL SECRETIONS WITH SPECIAL REFERENCE TO THE REGULATION OF GROWTH*

BURTON L. BAKER
University of Michigan Medical School, Ann Arbor, Michigan

A considerable body of evidence demonstrates that under certain experimental conditions body growth may be suppressed by treatment of animals with ACTH (Evans, Simpson, and Li, 1943), or pure C-11 oxygenated compounds (Wells and Kendall, 1940; Ingle, Sheppard, Oberle, and Kuizenga, 1946). It is to be expected that those tissues of the body which normally proliferate at a rapid rate would reflect most dramatically such general retardation in rate of growth. The purpose of this paper is to bring together some of these evidences of growth inhibition. Attention will be focused first on the effects of treatment of rats with adrenocorticotropin, contrasting these changes when possible to those which occur after adrenalectomy. This contrast is of interest because the occurrence of overgrowth following adrenalectomy might indicate that some adrenal steroids act as a brake on other growth-promoting influences at a normal physiological level.

In these studies, the ACTH was prepared by Dr. C. H. Li, the experiments were conducted by Dr. D. J. Ingle, and the anatomical work has been done in our laboratory. It should be kept in mind that, on the basis of present information, Dr. Li's preparation of ACTH appears to stimulate the adrenal cortex of the rat to secrete members of the C-11 oxygenated group of steroids and it may be assumed that most of the anatomical effects are mediated by them. Thus, the rats under discussion exhibited hyperglycemia, glycosuria, and a negative nitrogen balance. There was no evidence of salt retention.

Inhibition of Growth by Parenteral Administration of ACTH

Cartilage and Bone. In the ends of long bones are two tissues in which growth may be observed, namely, cartilage and bone. When cartilage cells proliferate, they become arranged in high columns. Formation of bone is indicated by the presence of osteoblasts (Fig. 1). ACTH retards growth in both tissues (Becks, Simpson, Li, and Evans, 1944). After treatment, the epiphyseal cartilage of adult rats is thinner, proliferation of the cartilage cells is reduced, as shown by the reduction in height of the cell columns and atrophic condition of the cells themselves (Fig. 2). Erosion of the cartilagenous matrix seems to be impaired. Likewise, the new formation of bone is retarded as indicated by the great reduction in number of active osteoblasts (Baker and Ingle, 1948).

---

* Aided by research grants from the Division of Research Grants and Fellowships, United States Public Health Service, and from The Upjohn Company.

Few studies are available of the epiphyses after adrenalectomy, but Wyman and Tum-Suden (1945) report that near the end of the second week after adrenalectomy in rats maintained on salt an increase in width of the epiphyseal cartilage can be detected.

Bone Marrow. ACTH causes atrophy of the red marrow of the tibia, this tissue being replaced by fat (Baker and Ingle, 1948). Similar changes of comparable intensity are not observed in the marrow of the vertebrae of the same animals. Therefore, it cannot be concluded that atrophy characterizes the bone marrow in general under these conditions.

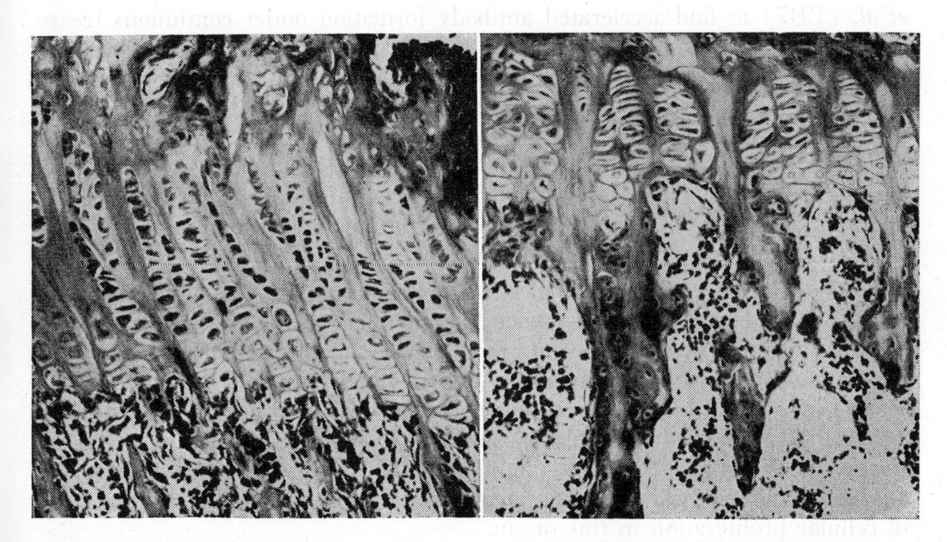

Fig. 1. Epiphyseal cartilage of tibia of adult rat.

Fig. 2. Epiphyseal cartilage of adult rat, 1 mg of ACTH daily for 21 days.

Most reports of the effect of adrenalectomy on the blood picture and bone marrow do not indicate hyperactivity of the bone marrow. Nevertheless, although an anemia follows adrenalectomy in rats, in animals maintained on 1% NaCl, Crafts (1941) finds that this condition is accompanied by normal or accelerated erythropoiesis since the percentage of reticulocytes is elevated and ultimately the hemoglobin content and erythrocyte count return toward normal values.

Lymphoid Tissue. The capacity of ACTH and C-11 oxygenated compounds to induce lymphoid atrophy is well established, and, largely as a result of the important work of Dougherty and White (1945), dissolution of lymphocytes is regarded as the immediate cause of this atrophy. Our observations suggest that continuous breakdown of lymphocytes under adrenal stimulation may not be the sole explanation for the maintenance of lymphoid atrophy. Thus, treatment of adult male rats with 3 mg daily of ACTH for 21 days results in an almost complete removal of lymphocytes from the thymus in many cases. However, in these same animals, although the lymphoid tissue in the spleen and lymph nodes shows some atrophy, many apparently normal lymphocytes are still present, the spleen in particular exhibiting well-formed Malpighian corpuscles. The point of particular importance here is that in the spleen and lymph nodes the rather normal nuclear structure of lymphocytes would indicate that they are not undergoing rapid dissolution. It should be noted that most previous work has been concerned primarily with the acute effects of adrenal substances on lymphocytes. In so far as the author is aware, maintenance of accelerated lymphocytic dissolution by prolonged administration of these hormones has not been demonstrated. In fact, failure to do so is indicated by the inability of Eisen *et al.* (1947) to find accelerated antibody formation under continuous treatment with adrenal steroids. One is thus brought to the expectation that inhibition of cell growth and proliferation may be one causative factor involved in the lymphoid atrophy which follows treatment with C-11 oxygenated steroids or ACTH. However, we must hasten to add that practically complete depletion of the splenic lymphocytes does follow more intense adrenocortical stimulation as is illustrated by a rat subjected to daily treatment with 6 mg of ACTH for 21 days by continuous injection superimposed upon the additional stress of immobilization.

Also, since seemingly normal lymphoid tissue is present in the spleen of ACTH-treated rats at a time when the thymus has undergone practically complete atrophy, it might be inferred that the thymus is more sensitive to the action of adrenal compounds than are the other lymphoid organs. When viewed in the light of the recent demonstration of Andreasen and Christensen (1949) that more rapid cellular proliferation occurs in the thymus than in the spleen or lymph nodes, one might infer that the greater ease with which adrenal compounds cause thymic atrophy is related to a higher normal rate of cellular proliferation in this organ.

Although there is some disagreement on the point (Stoerck, 1944) several investigators are agreed that adrenalectomy results in lymphoid hypertrophy, at least of the thymus (Jaffe, 1924; Reinhardt and Holmes, 1940).

Skin. In the skin are three epithelial structures which exhibit significant cellular proliferation, namely, hair bulbs, epidermis, and sebaceous glands. In each case, this proliferation serves to replace the products of cell metamorphosis.

Treatment of rats with 1 to 3 mg daily of ACTH for 21 days causes a nearly complete cessation of hair growth, this effect becoming evident during the third week of treatment (Figs. 3 and 4) (Baker *et al.*, 1948). Similar general inhibition in growth of hair is induced by systemic treatment with desoxycorticosterone acetate (Ralli and Graef, 1945) or adrenocortical extracts (Ingle and Baker, unpublished). Conversely, adrenalectomy accelerates the rate of hair growth (Butcher and Richards, 1939; Ralli and Graef, 1943). Treatment with ACTH results in thinning of the epidermis. On the other hand, thickening of the epidermis due to adrenalectomy has not been reported and has not occurred in any of our adrenalectomized animals.

Thus, there are 3 tissues, namely, lymphoid tissue, hair, and the epiphyseal cartilage of the tibia, the growth of which may be suppressed by hyperactivity of the adrenal cortex and which show a tendency toward overgrowth after ablation of the adrenal glands. It is probable that post-adrenalectomy overgrowth is a specific response of only certain tissues and does not reflect a general situation in the body. Certainly much evidence does not fit into this pattern, such as the retarded rate of liver regeneration (Berman, Sylvester, Hay, and Selye, 1947) and general body growth which results from adrenalectomy (Hartman and Thorn, 1930).

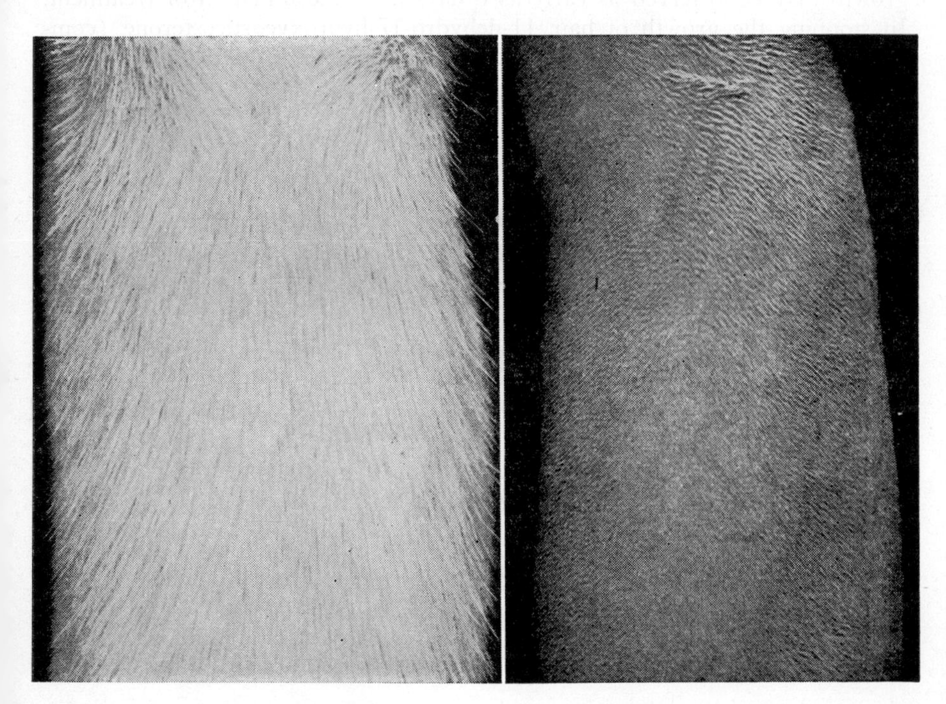

FIG. 3. Hair growth during third week of experiment.

FIG. 4. Hair growth during third week of treatment with 3 mg daily of ACTH.

INHIBITION OF GROWTH BY LOCAL ACTION OF ADRENAL STEROIDS

Work with the parenteral administration of ACTH, adrenal extracts, and pure compounds raises certain pertinent questions with respect to their inhibiting effects on growth: (1) Do these steroids act by inhibiting the secretion of growth-promoting substances by some other gland such as the anterior hypophysis or thyroid? (2) As has been postulated by others, is it true that growth is inhibited by adrenocortical secretions through a primary modification of liver metabolism?

Hair. In collaboration with Dr. Wayne L. Whitaker of our department we have undertaken to study these problems by observing the local growth-inhibiting action of adrenal compounds when applied directly to the skin. In these experiments the hormones are applied caudal to the right ear. If treatment with concentrated adrenal extract is continued long enough, atrophy of most of the major components of the skin may be induced in the area of treatment. These changes include, in the order of their appearance, cessation of hair growth as indicated by distortion of the normally symmetrical pattern of hair growth, thinning of the epidermis (Whitaker and Baker, 1948; Baker and Whitaker, 1948), and atrophy of the sebaceous glands. Thus, all the epithelial structures of the skin undergo regression. Inhibition of hair growth may be observed as early as 8 days after the initiation of treatment. In stopping the growth of hair 11-dehydro-17-hydroxycorticosterone (compound E) and 17-hydroxycorticosterone (compound F) are highly potent agents.

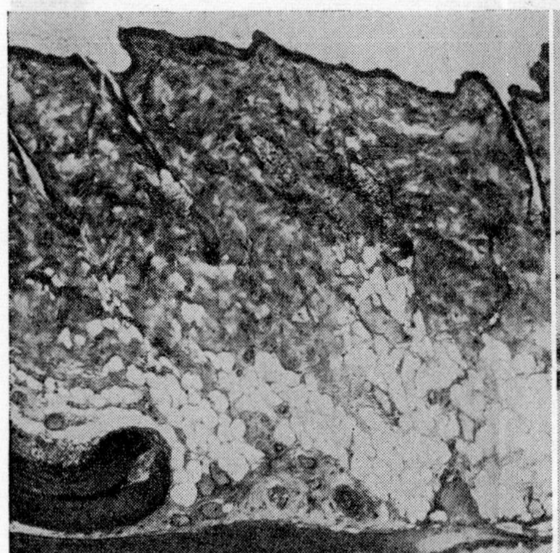

FIG. 5. Skin of left side of neck (nontreated).

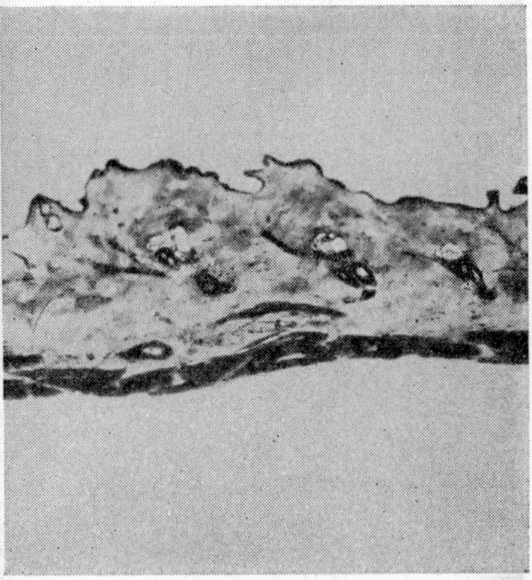

FIG. 6. Skin of right side of neck of same rat treated locally with 0.1 cc of hog adrenal extract daily for 112 days.*

In view of the current interest in the relationship of adrenal compounds to diseases involving the connective tissues, it is pertinent that an extreme thinning of the dermis may be induced by local hormone action (Figs. 5 and 6). The effect seems to be centered chiefly in the collagenous fibers which lose their sharp outline and appear to be fused into a compact mass. The elastic fibers are spared, being far more concentrated following treatment. Cellularity of the connective tissue is reduced. As revealed by the Hotchkiss periodic acid-leucofuchsin procedure, the glycoprotein basement membranes remain prominent. Thus, another characteristic of Cushing's syndrome, thin skin, may be reproduced by the local action of adrenal compounds.

Of particular interest also, since it indicates that adrenal compounds may mobilize fat from depots by direct action on the fat cell, is the complete disappearance of subcutaneous fat which occurs in many cases in the area of treatment.

---

* One cubic centimeter equivalent to 1 mg of 11-dehydro-17-hydroxycorticosterone by liver glycogen test.

We have attempted to inhibit the rate of wound repair by cutting out circular pieces of skin in symmetrical areas on the right and left sides of the neck after inhibition of hair growth had been established on the right side. Then, application of the extract to the right wound was continued. Although closure of the treated wound lags behind that of the non-treated one, still there is not much difference in the time of complete closure (Fig. 7). It may be concluded that the stimulus to growth caused by cutting the skin cannot be overcome by the local application of adrenal compounds under the conditions of our experiments.

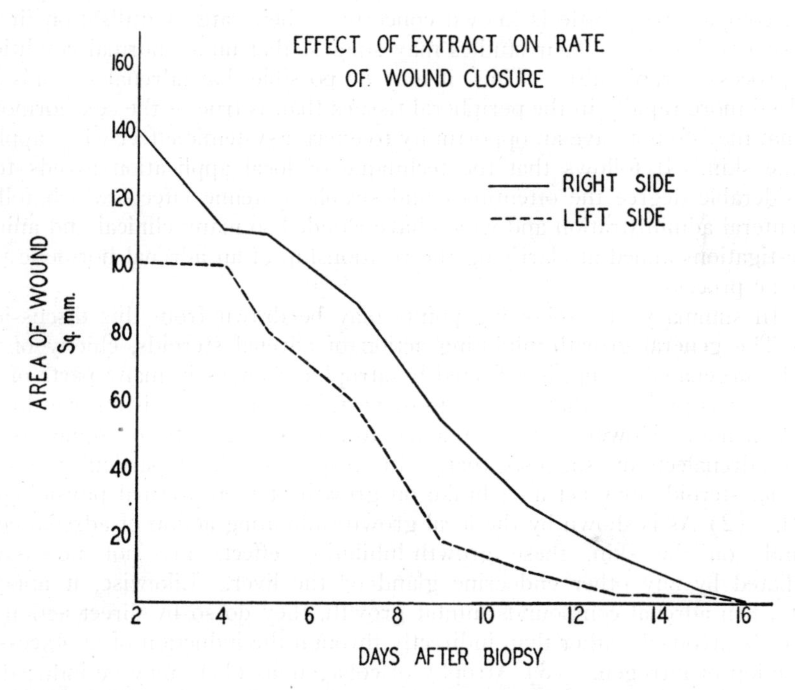

FIG. 7. Comparative rate of closure of wounds of right and left sides of neck, the right side being treated daily with adrenal extract equivalent to 100 μg of 11-dehydro-17-hydroxycorticosterone. Average of measurements on 14 rats.

The local action of adrenal hormones on the skin seems to place them in a unique position among the steroid hormones. We have not found evidence as yet of general systemic action when the hormones are applied to the skin although more delicate tests or higher dosages may reveal it. Growth of hair on the side of the neck opposite to the area of treatment continues in normal fashion, the inhibitory effects being limited precisely to the area of application. There is no reduction in thymic weight in these anmials. This lack of systemic effect is in direct contrast to the action of androgens or estrogens when they are applied to the skin. Sex hormones are so effective systemically when applied to the skin that one group of reviewers has concluded:

In general, it can be stated that provided a volatile solvent vehicle is used, the percutaneous and subcutaneous administration of sex hormones produce effects which are quantitatively of the same order. (Calvery, Draize, and Laug, 1946).

What is the reason for this difference in action? On the basis of present knowledge it would seem most likely that a fundamental difference in the rate or locus of utilization or inactivation between the adrenal and sex steroids may exist. Although it is well established that under conditions of stress, adrenal hormones are utilized in the peripheral tissues at an exceedingly rapid rate, comparatively little is known concerning their rate of utilization in the non-stressed animal. Our studies may suggest that under normal conditions the process is rapid also. Thus, it may be possible that adrenal steroids are utilized more rapidly in the peripheral tissues than is true of the sex hormones so that they do not have an opportunity to exert a systemic effect when applied to the skin. It follows that the technique of local application avoids to a considerable degree the oftentimes undesirable systemic effects which follow parenteral administration and which have clouded so many clinical and animal investigations aimed at clarifying the relationship of an adrenal hormone to a specific process.

In summary, the following points may be drawn from this discussion: (1) The general growth-inhibiting action of adrenal steroids, chiefly of the C-11 oxygenated group, is reflected by atrophic changes in many parts of the body. It is probable that, for the most part, such inhibition is an overdosage phenomenon. However, the tendency toward overgrowth of some tissues after adrenalectomy suggests that, with respect to these particular tissues, adrenal steroids may act as a brake on growth at their normal physiological level. (2) As is shown by the local growth-inhibiting action of adrenal compounds on the skin, these growth-inhibiting effects are not necessarily mediated by any other endocrine gland or the liver. Likewise, it appears that when adrenal compounds inhibit growth, they do so by direct action on the cells involved, rather than indirectly through the induction of an excessive excretion of nitrogen. (3) Atrophy of collagenous fibers may be induced by direct action of adrenal compounds. (4) Adrenal steroids differ from sex hormones in causing local inhibition of growth, without being systemically active when applied to the skin. (5) Cutting the skin in an area of inhibited growth resulting from the direct application of adrenal steroids causes the re-initiation of growth.

REFERENCES

ANDREASEN, E., and CHRISTENSEN, S., Rate of mitotic activity in the lymphoid organs of the rat. *Anat. Record,* **103,** 401-412 (1949).

BAKER, B. L., and INGLE, D. J., Growth inhibition in bone and bone marrow following treatment with adrenocorticotropin (ACTH). *Endocrinology,* **43,** 422-429 (1948).

BAKER, B. L., and WHITAKER, W. L., Growth inhibition in the skin following direct application of adrenal cortical preparations. *Anat. Record,* **102,** 333-348 (1948).

BAKER, B. L., INGLE, D. J., LI, C. H., and EVANS, H. M., Growth inhibition in the skin induced by parenteral administration of adrenocorticotropin. *Anat. Record,* **102,** 313-332 (1948).

BECKS, H., SIMPSON, M. E., LI, C. H., and EVANS, H. M., Effects of adrenocorticotropic hormone (ACTH) on the osseous system on normal rats. *Endocrinology,* **34,** 305-310 (1944).

BERMAN, D., SYLVESTER, M., HAY, E. C., and SELYE, H., The adrenal and early hepatic regeneration. *Endocrinology,* **41,** 258-264 (1947).

BUTCHER, E. O., and RICHARDS, R. A., The relation of the adrenals to the retarded hair growth in underfed albino rats. *Endocrinology,* **25,** 787-792 (1939).

CALVERY, H. O., DRAIZE, J. H., and LAUG, E. P., The metabolism and permeability of normal skin. *Physiol. Revs.,* **26,** 495-540 (1946).

CRAFTS, R. C., The effects of endocrines on the formed elements of the blood. I. The effects of hypophysectomy, thyroidectomy and adrenalectomy on the blood of the adult female rat. *Endocrinology,* **29,** 596-605 (1941).

DOUGHERTY, T. F., and WHITE, A., Functional alterations in lymphoid tissue induced by adrenal cortical secretion. *Am. J. Anat.,* **77,** 81-116 (1945).

EISEN, H. N., MAYER, M. M., MOORE, D. H., TARR, R., and STOERCK, H. C., Failure of adrenal cortical activity to influence circulating antibodies and gamma globulin. *Proc. Soc. Exptl. Biol. Med.,* **65,** 301-306 (1947).

EVANS, H. M., SIMPSON, M. E., and LI, C. H., Inhibiting effect of adrenocorticotropic hormone on the growth of male rats. *Endocrinology,* **33,** 237-238 (1943).

HARTMAN, F. A., and THORN, G. W., A biological method for the assay of cortin. *Proc. Soc. Exptl. Biol. Med.,* **28,** 94-95 (1930).

INGLE, D. J., HIGGINS, S. M., and KENDALL, E. C., Atrophy of the adrenal cortex in the rat produced by administration of large amounts of cortin. *Anat. Record,* **71,** 363-372 (1938).

INGLE, D. J., SHEPPARD, R., OBERLE, E. A., and KUIZENGA, M. H., A comparison of the acute effects of corticosterone and 17-hydroxycorticosterone on body weight and the urinary excretion of sodium chloride, potassium, nitrogen and glucose in the normal rat. *Endocrinology,* **39,** 52-57 (1946).

JAFFE, H. L., The influence of the suprarenal gland on the thymus. I. Regeneration of the thymus following double suprarenalectomy in the rat. *J. Exptl. Med.,* **40,** 325-342 (1924).

RALLI, E. P., and GRAEF, I., Stimulating effect of adrenalectomy on hair growth and melanin deposition in black rats fed diets adequate and deficient in the filtrate factors of vitamine B. *Endocrinology,* **32,** 1-12 (1943).

RALLI, E. P., and GRAEF, I., The effects of the synthetic and natural hormones of the adrenal cortex on melanin deposition in adrenalectomized black rats fed diets adequate and deficient in the filtrate factors of vitamin B. *Endocrinology,* **37,** 252-261 (1945).

REINHARDT, W. O., and HOLMES, R. O., Thymus and lymph nodes following adrenalectomy and maintenance with NaCl in the rat. *Proc. Soc. Exptl. Biol. Med.,* **45,** 267-270 (1940).

STOERCK, H. C., Thymus weight in relation to body weight in castrated and in adrenalectomized rats. *Endocrinology,* **34,** 329-334 (1944).

WELLS, B. B., and KENDALL, E. C., A qualitative difference in the effect of compounds separated from the adrenal cortex on distribution of electrolytes and on atrophy of the adrenal and thymus glands of rats. *Proc. Staff Meetings Mayo Clinic,* **15,** 133-139 (1940).

WHITAKER, W. L., and BAKER, B. L., Inhibition of hair growth by the percutaneous application of certain adrenal cortical preparations. *Science,* **108,** 207-209 (1948).

WYMAN, L. C., and TUM-SUDEN, C., The effect of adrenalectomy on the epiphyseal cartilage of the rat. *Endocrinology,* **36,** 340-346 (1945).

PRELIMINARY OBSERVATIONS ON THE RELATION OF THE ADRENAL CORTEX TO ELECTROLYTE METABOLISM IN THE RAT

ROY O. GREEP

School of Dental Medicine, Harvard University, Boston, Massachusetts

Many of the metabolic deficiences which develop after removal of the adrenal glands resemble those seen after hypophysectomy. These similarities are understandable, since the cortex is largely dependent on the pituitary. It is the dissimilarities in the effects of adrenalectomy as compared to hypophysectomy, however, which stimulated the work to be reported here.

The effects of adrenalectomy differ from those of hypophysectomy in two outstanding ways: (1) When the adrenals are removed, the rat succumbs quickly (Cowie, 1949), whereas the animal lacking a pituitary can survive for an indefinite period under good laboratory conditions. (2) The adrenalectomized rat has an increased appetite for sodium chloride and may survive indefinitely if its salt requirement is met; the hypophysectomized animal has no altered salt requirements. These differences in postoperative history are surely reflections of a basic dichotomy in the metabolic sequelae to the removal of these two endocrine glands.

After hypophysectomy the adrenal glands shrink to less than half their former size. All this loss of weight is due to reduction in the size of the cortex. It has been demonstrated for many species, however, that the zona glomerulosa, the outer border of the cortex, fails to undergo atrophy after removal of the pituitary (dog, Houssay and Sammartino, 1933; rat, Deane and Greep, 1946; mouse, Jones, 1949; guinea pig, Schweizer and Long, 1950). In all these species, the glomerulosa appears healthy, even long after the removal of the pituitary. Closer histochemical studies have also shown that the cells of this zone, in the rat, retain all the characteristics of actively secreting cortical cells. Because the hypophysectomized animal does not require added salt and because the glomerulosa remains normal after hypophysectomy, it has been suggested that the glomerulosa, in the rat at least, is secreting a salt-regulating hormone in the absence of the pituitary.

I need not review here all the cytochemical evidence which has been brought forward in support of this contention, as that will be found in the paper by Deane in this volume. From the physiological point of view, it is pertinent to recall that a steroid, desoxycorticosterone, which is markedly effective in regulating the salt metabolism of adrenalectomized rats, does cause the zona glomerulosa to undergo selective atrophy when given to intact rats. Furthermore, when this substance is administered to hypophysectomized rats, after the adrenals have essentially ceased to lose weight, a further significant reduction in adrenal size takes place as a result of an induced atrophy of the remaining zona glomerulosa (Sarason, 1943; Greep and Deane, 1947). These results plainly suggest that the outer border of the rat's adrenal cortex is in some way related to electrolyte metabolism. If that be true, then this same tissue should be responsive to alterations in the circulating electrolytes induced by dietary means. The evidence obtained by this approach has only served to bolster the original hypothesis (Sarason, 1943; Deane, Shaw, and Greep, 1949). In a recent clinical study Leaf and Couter (1949) have concluded that the desoxycorticosterone-like activity of the adrenal cortex is depressed by sodium administration.

A further examination of this hypothesis is in progress. I am presenting here some of the findings thus far obtained.

The primary problem is to obtain conclusive evidence that the adrenal glands of the hypophysectomized rat are serving a vital function. The obvious approach here is to remove the adrenals of hypophysectomized rats and determine whether the typical symptoms of acute adrenal insufficiency develop. In order to carry out this test in a critical fashion, it is necessary to have the assurance that the animals will survive for a long time after the pituitary is removed. This, I believe, has been achieved through the use of a purified diet which contains all the known essential dietary ingredients, is high in caloric content, and is palatable (Shaw and Greep, 1949). Fortified with this information, I. Chester Jones and I removed the adrenals from rats that had been hypophysectomized for varying periods up to 180 days. The two adrenals were removed in one stage. The operation was done rapidly, with a minimum of trauma and a negligible blood loss. The animals were kept in single cages in a well-heated room and provided with food *ad libitum*. When 15 such rats were given tap water, 1 died on the second day after adrenalectomy, 1 on the third, 7 on the fourth, 5 on the fifth, and the last one on the sixth day. The terminal symptoms were completely similar to those seen in non-hypophysectomized rats after adrenalectomy. The fall in body temperature, the shivering, the characteristic hunched posture (in which they often died), the convulsions, diarrhea, and precipitous decline were unmistakably those of adrenal insufficiency. These agonal symptoms bear no resemblance to those through which the ordinary hypophysectomized animal passes.

It was a matter of some surprise that these doubly operated animals could not be maintained with 1% NaCl in the drinking water. Twenty-three rats in one group were adrenalectomized after hypophysectomy and given 1% saline; 5 died within 6 days. The NaCl in the drinking water was then raised to 2%. Seventeen rats died in the next 2 days. The added salt was clearly of no benefit; the increased death rate may not, however, have been due to the added salt. Eight doubly operated rats in another group were given 0.6% NaCl; these did not die until the fourteenth to the twenty-third day after adrenalectomy. This slight improvement in survival time suggests that further work along this line of regulating salt intake may disclose the proper salt requirement of these doubly operated animals. We have tried a drinking fluid made up of 0.45% sodium chloride and 0.45% sodium citrate with little success. It was felt that with the hypotension and reduced renal plasma flow which attends the hypophysectomized state and is, no doubt, intensified by later adrenalectomy, the kidneys may not be able to concentrate the chloride ion. That this might not be the case is indicated by the fact that this same solution was given to hypophysectomized rats for 30 days without ill effects.

If the adrenals of the hypophysectomized rat do secrete a hormone having salt-regulating properties, then the replacement of that activity with desoxycorticosterone might be expected to afford an equivalent protection. Four rats adrenalectomized 35 days after pituitary ablation were given 2.5 mg desoxycorticosterone daily beginning immediately after the second operation. After 13 days of treatment, there had been no deaths. Since the animals were eating well and there was no evident change in their physical condition, the desoxycorticosterone was discontinued. Fourteen days later 2 were dead, and another died the next day. It is quite probable that in the 2 weeks between cessation of desoxycorticosterone injections and the first death, the oil at the injection sites continued to yield diminishing amounts of hormone for about one week, so that the survival following cessation of effective desoxy-

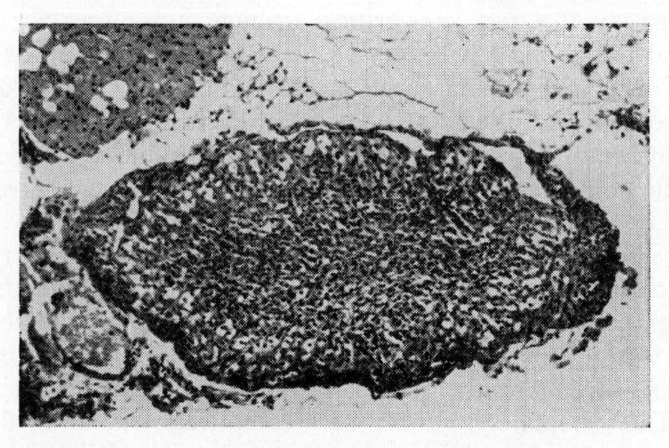

FIG. 1. Adrenal remnant in a rat surviving hypophysectomy and adrenalectomy.

corticosterone activity is not different from that following adrenalectomy. The fourth rat in this group continued to live and was killed 110 days after adrenalectomy (97 days after the hormone treatment was discontinued). The retroperitoneal fat and lymphatic tissues lying forward of the kidneys were fixed and sectioned serially. In this we found one small adrenal rest (Fig. 1). The structure of this body was of unusual interest. There was a small core of atrophic cells surrounded by a wide outer border composed of smaller and vacuolated cells. It was not possible to identify the central atrophic cells as remnants of zona fasciculata, but their resemblance to such cells in the adrenals of long-term hypophysectomized rats was striking. The outer border was, in all probability, a zona glomerulosa. It unquestionably appears hyperfunctional and closely resembles the zona glomerulosa of rats kept on a sodium-deficient diet for several weeks (Deane, Shaw, and Greep, 1949).

In view of these various observations, it seemed desirable to investigate the electrolyte balance of these doubly operated rats in comparison to normal rats and to rats with only the pituitary removed. The intake of sodium and potassium was calculated from the amount of food consumed. The output of these electrolytes in the urine was measured. The determinations were made with a Beckman flame photometer.

Two long-term hypophysectomized rats and 2 young intact controls were studied over several weeks (Greep and Roby, unpublished). The intact animals seldom varied from a positive sodium and positive potassium balance. Some retention of both elements was to be expected, since these animals were still growing. The hypophysectomized rats shifted irregularly from a slightly positive to a slightly negative balance but, all in all, displayed about an even balance, which is in keeping with an unaltered salt requirement and maintenance of a uniform body weight. In terms of total daily turnover of electrolytes per 100 g body weight, the values for the operated rats ran roughly from one-third to one-half that of their intact controls.

Another group of 5 rats was used to follow the electrolyte intake-output pattern during the normal state, after hypophysectomy, and after subsequent adrenalectomy. The results are given in Table 1. It will be seen that these

TABLE 1

Average Intake and Urinary Excretion of Electrolytes by a Group of 5 Male Rats before and after Hypophysectomy and after Subsequent Adrenalectomy.

The values are expressed in milliequivalents per 24 hours per 100 g body weight. The pituitaries were removed on the fourth day of the experiment and the adrenals on the tenth day.

Day of Experiment	Sodium		Potassium	
	Intake	Urine output	Intake	Urine output
	Intact			
1st	1.20	0.814	0.750	0.744
2nd	1.50	0.965	0.960	0.672
3rd	1.28	0.730	0.800	0.541
	Hypophysectomized			
5th	0.409	0.266	0.258	0.375
6th	0.575	0.351	0.360	0.512
7th	0.505	0.295	0.316	0.536
	Hypophysectomized-Adrenalectomized			
12th	0.571	0.454	0.357	0.439
13th	0.555	0.350	0.347	0.445
14th	0.277	0.326	0.173	0.326

rats were in positive sodium balance and positive potassium balance while intact. After hypophysectomy there was no change in the sodium balance, but the potassium shifted to a negative balance. The latter result may be a reflection of an excess tissue breakdown in the first few days after hypophysectomy, as this distinct negative balance does not characterize the long-term hypophysectomized rat. The adrenals were removed 1 week after hypophysectomy, and no change occurred in the electrolyte balances nor in the absolute amounts of sodium and potassium consumed or excreted. All the rats died within 6 days of adrenalectomy, yet these measurements of their electrolyte metabolism did not furnish a clue as the cause of death. These data are, however, of only preliminary nature, and it is possible that with modifications and refinement of this approach subtle changes may be found. It seems fairly certain, however, that the striking changes in electrolyte excretion which are well known to occur in plain adrenalectomized rats do not supervene if the pituitary is first removed. As a check on our methods, the excretion of the electrolytes was measured before and after adrenalectomy, leaving the pituitary intact. The average pre-operative excretion of sodium by 4 rats was 0.746 milliequivalent per 100 g body weight per 24 hours and rose to 1.351 milliequivalents on the fifth postoperative day; simultaneously the potassium excretion dropped from 0.746 milliequivalent per 100 g body weight per 24 hours to 0.49 milliequivalent.

An alternate means of exploring the extent to which the electrolyte-regulating function of the adrenal cortex in the rat is independent of the pituitary is to observe the effect of adrenocorticotropin (ACTH) on electrolyte excretion. Ingle, Li, and Evans (1946) treated intact rats with 7 mg ACTH daily and found no change in sodium or chloride excretion in spite of considerable adrenal enlargement; although the excretion of potassium was often increased, this appeared to accompany a rise in urinary NPN and may have resulted from the additional tissue breakdown. Bergner and Deane (1949) also treated intact rats with ACTH and found no change in the excretion of sodium or chloride and, even though the urinary NPN was increased, no accompanying change in potassium was noted. Roby and I gave 6 mg ACTH daily to normal and long-term hypophysectomized rats on several occasions for 24, 48, or 96 hours and were unable to detect a change in sodium or potassium balance and in the absolute amounts of these substances excreted.

In summary, the adrenal cortex has been shown to serve a vital function in the hypophysectomized rat. The critical deficiency which follows removal of the adrenals in this circumstance has not been determined. Salt therapy in the conventional manner (1% NaCl) was not successful in maintaining life after adrenalectomy, although a lesser amount was of some benefit. Desoxycorticosterone acetate was more effective than salt in this regard.

A study of the intake and output of sodium and potassium has shown that the hypophysectomized rat is able to maintain an effective electrolyte balance, and this was not disturbed during the brief survival following removal of the adrenals. Such findings are not of crucial significance, since they do not preclude possible critical changes in the blood levels of these electrolytes or of shifts in their compartmental distribution.

REFERENCES

Bergner, G. E., and Deane, H. W., Effects of pituitary adrenocorticotropic hormone on the intact rat, with special reference to cytochemical changes in the adrenal cortex. *Endocrinology,* **43,** 240 (1948).

Cowie, A. T., The influence of age and sex on the life span of adrenalectomized rats. *J. Endocrinol.,* **6,** 94 (1949).

Deane, H. W., and Greep, R. O., Morphological and histochemical study of rat's adrenal cortex after hypophysectomy, with comments on liver. *Am. J. Anat.,* **79,** 117 (1946).

DEANE, H. W., SHAW, J. H., and GREEP, R. O., The effect of altered sodium or potassium intake on the width and cytochemistry of the zona glomerulosa of the rat's adrenal cortex. *Endocrinology,* **43,** 133 (1949).

GREEP, R. O., and DEANE, H. W., Cytochemical evidence for the cessation of hormone production in the zona glomerulosa of the rat's adrenal cortex after prolonged treatment with desoxycorticosterone acetate. *Endocrinology,* **40,** 417 (1947).

HOUSSAY, B. A., and SAMMARTINO, R., Histología de las suprarrenales de los perros hipofisoprivos. *Rev. soc. argentina biol.,* **9,** 209 (1933).

INGLE, D. W., LI, C. H., and EVANS, H. W., The effect of adrenocorticotrophic hormone on the urinary excretion of sodium, chloride, potassium, nitrogen and glucose in normal rats. *Endocrinology,* **39,** 32 (1946).

JONES, I. C., The relationship of the mouse adrenal cortex to the pituitary. *Endocrinology,* **45,** 514 (1949).

LEAF, A., and COUTER, W. T., Evidence that renal sodium excretion by normal human subjects is regulated by adrenal cortex activity. *J. Clin. Invest.,* **28,** 1067 (1949).

SARASON, E. L., Morphologic changes in rat's adrenal cortex under various experimental conditions. *Arch. Path.,* **35,** 373 (1943).

SCHWEIZER, M., and LONG, M. E., Partial maintenance of the adrenal cortex by anterior pituitary grafts in fed and starved guinea pigs. *Endocrinology,* **46,** 191 (1950).

SHAW, J. H., and GREEP, R. O., Relationship of diet to the duration of survival, body weight, and composition of hypophysectomized rats. *Endocrinology,* **44,** 520 (1949).

THE ADRENAL CORTEX IN WATER METABOLISM*

ROBERT GAUNT, JAMES H. BIRNIE, W. R. BOSS, W. J. EVERSOLE, AND C. M. OSBORN

Syracuse University, Syracuse, New York

The role of adrenal cortical hormones in water and electrolyte metabolism is a complex one. It is made the more confusing by the fact that the specific actions of these hormones vary greatly depending upon the setting of the physiological stage on which they operate. In the short time available, we shall attempt to discuss just one aspect of the subject, namely, the mechanisms by which cortical hormones affect water diuresis. This problem resolves itself into two specific questions: First, why is it that either animals or men without normal adrenals are unable to show a normal diuretic response to administered water? The fact that they cannot do so is the basis for part of the Robinson-Power-Kepler (1941) water test for Addison's disease. Secondly, why is it that when animals are overdosed with cortical hormones, the rate of excretion of ingested fluid can be accelerated above normal? In other words, by what means are the cortical hormones diuretic agents? For brevity, we will limit ourselves largely to the work done on the rat since that is the form with which we have had most experience. A more comprehensive review of this subject has recently been published (Gaunt, Birnie, and Eversole, 1949).

Failure of Diuretic Mechanisms in Adrenal Insufficiency

There are a number of factors which could or do contribute to the fact that animals with even a mild adrenal insufficiency exhibit a greatly retarded diuretic response to the administration of water. Among these are the following:

1. When fluid is given by mouth there is a delayed stomach-emptying time and decreased rate of intestinal absorption of water. As a practical matter, this factor is of little consequence as can be shown readily by the fact that when intestinal fluid is eventually absorbed, it is not rapidly excreted, and if fluid is introduced parenterally it is excreted little if any better than that given by mouth (Gaunt, 1944; Joseph *et al.*, 1944).
2. There are other extrarenal influences which contribute to the abnormal disposal of fluid after adrenalectomy. They seem to affect the internal distribution of absorbed fluid. Although the existence of such influences has been brought into sharp focus in nephrectomized animals, little is known as to their specific nature (Birnie *et al.*, 1948).

* This work was aided by grants from the National Heart Institute, U. S. Public Health Service, and Ciba Pharmaceutical Products, Inc.

3. Renal factors are, as might be expected, of great importance. A decreased renal plasma flow and glomerular filtration rate has been seen by many workers, e.g., Talbott *et al.*, 1942; White *et al.*, 1947; Waterhouse and Keutmann, 1948; Sanderson, 1946-48. In the rat the reduction of the filtration rate may be in the order of 25 to 30%. If water is given in amounts sufficient to produce beginning symptoms of water intoxication, the filtration rate descends to less than half normal. Even so, this does not seem to be the primary factor in inhibiting water diuresis (Boss *et al.*, 1950).

4. A renal factor of greater magnitude is that of an increased tubular reabsorption of water. Reabsorption may increase in hydrated animals from 60 to 90% above normal, and studies with replacement therapies have shown that the actual rate of diuresis in adrenalectomized animals is more closely correlated with the rate of tubular reabsorption than with any other variable studied (Boss *et al.*, 1950; Lotspeich, 1949; Roemmelt *et al.*, 1949).

The Cause of Increased Tubular Reabsorption after Adrenalectomy. If it is granted that an increased tubular reabsorption is an essential feature in the failure of water diuresis after adrenalectomy, then the question arises as to what causes such changes in tubular function. It seemed to us most logical to look for the appearance of a humoral stimulant of reabsorption after adrenalectomy, since normally urine is concentrated under the influence of a humoral agent, the pituitary antidiuretic hormone (ADH). Several years ago Martin *et al.* (1939) found that the urine of adrenalectomized cats contained an antidiuretic substance. We turned to an examination of the blood serum and found that rat serum when tested in hydrated rats contained a labile, chloruretic, antidiuretic substance (ADS) which increased in amounts after adrenalectomy. Information obtained to date concerning this substance is as follows (Birnie *et al.*, 1949, 1950):

1. It reduces water excretion and increases chloride excretion.
2. If allowed to stand it disappears rapidly from fresh serum.
3. It is not detectable in the blood of hypophysectomized rats.
4. It is increased in the blood of adrenalectomized rats.
5. It acts by causing an increased tubular reabsorption of water despite a transient elevation of glomerular filtration rate.
6. It is active by subcutaneous injection but, like Pitressin, is more effective when given intraperitoneally.
7. Although blood for tests was generally drawn under ether anesthesia, the presence of antidiuretic activity is not dependent upon etherization.
8. It is inactivated by neutralized thioglycollic acid as in Pitressin.
9. It is inactivated by a cell-free liver extract (enzyme) as is Pitressin.
10. Its action is distinguishable from those of nonspecific irritating foreign proteins by the fact, among others, that foreign substances are not chloruretic.

These properties suggest that this substance is the circulating pituitary ADH which for some reason accumulates after adrenalectomy. But in view of the fact that so many antidiuretic agents have been reported in body fluids and tissues, we prefer to reserve judgment on the question of identity until further supporting evidence is available.

Preliminary studies, employing the hydrated rat test, indicate the presence of serum ADS in hamsters, rabbits, and dogs. Lloyd and Lobotsky (1949) have reported a somewhat similar material in human serum and have demonstrated a correlation between the amounts present and the level of corticoid excretion.

Cause of Accumulation of ADS after Adrenalectomy. The next question that might be raised is, why does an ADS accumulate after adrenalectomy? No certain answer is available but some tentative clues have been obtained which serve as a basis for further work.

The first possibility is that the neurohypophysis is hyperactive after adrenalectomy and thus maintains higher-than-normal levels of ADH in the blood. This possibility has not been eliminated but there are theoretical reasons (Gaunt, 1944) and some direct evidence (Gersh and Grollman, 1939) for doubting that there would be a hypersecretion of ADH in adrenal insufficiency. Since the major stimulant to ADH secretion is thought to be a rise in the osmotic pressure of plasma (Chambers *et al.*, 1945; Verney, 1946), and since in adrenal insufficiency, particularly after water administration (Swingle *et al.*, 1937; Gaunt, 1944), the major osmotic constituents of plasma are reduced in concentration, an increased output of ADH would hardly be expected.

On the other hand, there is the possibility of a decreased destruction of antidiuretic agents in adrenal insufficiency. Evidence now available indicates that the liver is involved in the inactivation of ADH. This inactivation is probably due to an enzyme system, one not limited to hepatic tissue but found in highest concentration in this tissue (Eversole *et al.*, 1949). The active agent appears in cell-free extracts. It inactivates both posterior pituitary extracts and the ADS of rat serum, although it may not be specific for antidiuretic hormones. It is of particular interest for the present discussion that this inactivating agent in the liver decreases in amount after adrenalectomy (Birnie, 1950). Therefore, as a working hypothesis, we are studying the possibility that the accumulation of ADS after adrenalectomy may be due to a decreased destruction, rather than overproduction of that substance.

The Accumulation of ADS Is not the Sole Cause of Antidiuresis in Adrenal Insufficiency. Whatever may be the role and significance of this ADS, its accumulation is not the sole cause of the inhibited water diuresis after adrenalectomy because as recent studies show:

1. Maintenance with salt or small doses of DCA or adrenal cortical extract will prevent accumulation of the ADS above normal levels, but will not restore normal water diuresis in adrenalectomized animals.

2. An inhibited diuretic response to water appears after adrenalectomy earlier than a detectable increase in ADS.

Hypersensitivity to Antidiuretic Substances. The factor accounting for failure of diuresis when the ADS is not elevated is probably a hypersensitivity to antidiuretic agents. This is manifested by an increased response to pituitary ADH (Corey *et al.,* 1939; Birnie *et al.,* 1950) and serum ADS after adrenalectomy. Therefore, as long as circulating antidiuretic substances are not decreased in adrenal insufficiency and the antagonistic diuretic cortical hormones are absent, a normal water diuresis would hardly be expected.

It would seem then that there is a series of influences released in the absence of cortical hormones which act more or less cumulatively to increase the renal tubular reabsorption of water and thus inhibit a full diuretic response to water.

Diuretic Action of Adrenal Cortical Hormones in Intact Animal

Whether the cortical hormones induce diuresis in intact animals, or fail to do so, depends upon the hormone preparation used, the water load, the salt load, and other variables suggested in Table 1.

TABLE 1

Conditions in Which Cortical Hormones Stimulate Diuresis in Intact Animals

Fluid	DCA	Adrenal Cortical Extract
Chronic Treatment		
Water *ad lib*	+	+
Saline *ad lib*	++	?
Acute Treatment		
Excess hydration	+	++
Mild hydration	+	+
Slight hydration	0	?
Mild dehydration	0	+
Normal saline	0	++
Mild hydration in newborn rats	?	++
Mild hydration in hypophysectomized rats	+	++

It has been established that the chronic administration of DCA leads to a diabetes insipidus-like syndrome (Kuhlman *et al.,* 1939; Ragan *et al.,* 1940). Ingle (1949) has found in unpublished observations that the same is true of adrenal cortical extract. If saline is used as a drinking solution or salt added otherwise, the effect of DCA is greatly increased as many workers have found, e.g., Harned and Nelson, 1943; Rice and Richter, 1943.

Another approach to the problem is that of giving fluid by stomach tube and observing the effect of acute overdosage with various hormones on the rate of urine flow. When huge doses of water were given by this means, i.e., in amounts expected to induce water intoxication, it was found that both DCA and cortical extract would increase the rate of water output. Whole extract, in fact, was dramatically effective, and its use made it almost impossible to produce lethal water intoxication (Gaunt, 1943).

Under other conditions, the results varied depending upon the degree of hydration, the hormone preparation used, etc., as follows (Osborn and Eversole, 1949; Sartorius and Pitts, 1949; Birnie, 1950). When mild hydration was produced by giving one or two standard diuretic doses of water, again both DCA and cortical extract elevated the diuretic rate above normal. With this level of hydration the effect, although highly consistent, was not of great magnitude (*ca.*+22% at 90 minutes). DCA increased urine volume little or none after the administration of a small dose of water (2 ml per 100 g body weight), or in mild dehydration (withholding water for 20 hours), or after normal saline administration (4 ml per 100 sq cm body surface). In all these conditions adrenal cortical extract did increase the diuretic rate, and particularly so after the administration of normal saline (+57% at 90 minutes).

Newborn rats normally show a very limited diuretic response to the ingestion of water (Heller, 1947; McCance and Wilkinson, 1947). This report has been confirmed and the additional observation made that cortical extract given with water will permit a diuretic response in the newborn (Osborn and LoCascio, 1950).

It is of incidental interest here that both DCA and cortical extract stimulate water loss in the mildly hydrated hypophysectomized rats, with the extract being much more effective than DCA.

Thus it is seen that DCA stimulates water diuresis when it is given chronically or when in acute experiments the water load is high. Cortical extract works under all conditions tried.

Several students of the DCA-induced diabetes insipidus-like syndrome think that the high water exchange is due primarily to an increased thirst resulting from the sodium-retaining action of this hormone. Thus polydipsia would be the cause of polyuria. At first glance, it might be thought that the weakness of DCA in enhancing the excretion of fluid given as normal saline in acute experiments is inconsistent with this theory. That is probably not true because it is well known that the salt retaining action of DCA is associated with an osmotic retention of water and in Addisonian patients may lead to edema. This action may nullify or override the diuretic properties of the hormone. When the chronic administration of DCA builds up a sufficient salt load to increase thirst in the normal animal, then this additional factor of thirst could become quantitatively the dominant one.

Such considerations will not explain the increased diuresis that follows the acute administration of water when cortical hormones are given. Here the additonal influence seems to be again the effect of these hormones on the tubular reabsorption of water. When mild hydration (2 hourly doses of 3 ml per 100 sq cm body surface) is produced in intact rats, both DCA and cortical extract increase the rate of urine flow without affecting the glomerular filtration rate (Table 2). The inference is, therefore, that the cortical

TABLE 2

Effect of Cortical Hormones on Hydrated Intact Animals

	Glomerular Filtration (ml/100 g/min.)	Per Cent Increase in Urine Flow
Normal	1.002	.....
DCA, 5 mg	0.90	+25
Cortical extract, 4 ml	0.99	+27

hormones inhibit the tubular reabsorption of water, just as, conversely, reabsorption is enhanced after adrenalectomy. The mechanisms by which this action is accomplished are not known, but the suggestion is obvious that they may act by antagonizing in some way the actions of pituitary ADH either directly or consequent to effects on sodium metabolism. Further studies to clarify these points are in progress.

Summary and Conclusions

Britton and co-workers (Corey, Silvette, and Britton, 1939) were primarily responsible for the theory that the hormones of the adrenal cortex and posterior pituitary exert antagonistic influences in the metabolism of both sodium salts and water. There is much to support this concept in the experiments briefly reviewed here. The influences which affect water metabolism, however, are numerous and their interactions devious. The high-lighting of any one factor will depend upon experimental design permitting its emphasis. Throughout all of this work, nevertheless, a consistent action of the hormones of the adrenal cortex, as present in whole gland extracts, is to stimulate diuresis. They accomplish this, clearly, by more than one means, but one consistent aspect of their action is to inhibit by direct or indirect ways the renal tubular reabsorption of water. In this respect the adrenal cortical and posterior pituitary hormones are clearly antagonistic agents, but the extent to which this antagonism constitutes a correlated and integrated physiological mechanism remains to be detemined.

REFERENCES

Birnie, J. H., *Federation Proc.,* **9,** 12 (1950).

Birnie, J. H., Eversole, W. J., and Gaunt, R., *Endocrinology,* **42,** 412 (1948).

Birnie, J. H., Jenkins, R., Eversole, W. J., and Gaunt, R., *Proc. Soc. Exptl. Biol. Med.,* **70,** 83 (1949).

Birnie, J. H., Eversole, W. J., Boss, W. R., Osborn, C. M., and Gaunt, R., *Endocrinology,* **47,** 1 (1950).

Boss, W. R., Birnie, J. H., and Gaunt, R., *Endocrinology,* **46,** 307 (1950).

Chambers, G. H., Melville, E. V., Hare, R. S., and Hare, K., *Am. J. Physiol.,* **144,** 311 (1945).

Corey, E. L., Silvette, H., and Britton, S. W., *Am. J. Physiol.,* **125,** 644 (1939).

Eversole, W. J., Birnie, J. H., and Gaunt, R., *Endocrinology,* **45,** 378 (1949).

Gaunt, R., *Proc. Soc. Exptl. Biol. Med.,* **54,** 19 (1943).

Gaunt, R., *Endocrinology,* **34,** 400 (1944).

Gaunt, R., Birnie, J. H., and Eversole, W. J., *Physiol. Revs.,* **29,** 281 (1949).

Gersh, I., and Grollman, A., *Am. J. Physiol.,* **125,** 66 (1939).

Harned, A. S., and Nelson, W. O., *Federation Proc.,* **2,** 19 (1943).

Heller, H., *J. Physiol.,* **106,** 245 (1947).

Ingle, D. J., personal communication, 1949.

Joseph, S., Schweizer, M., Ulmer, N. Z., and Gaunt, R., *Endocrinology,* **35,** 338 (1944).

Kuhlman, D., Ragan, C., Ferrebee, J. W., Atchley, D. W., and Loeb, R. F., *Science,* **90,** 496 (1939).

Lloyd, C. W., and Lobotsky, J., *Am. J. Med.,* **7,** 415 (1949).

Lotspeich, W. D., *Endocrinology,* **44,** 314 (1949).
Martin, S. J., Herrlich, H. C. and Fazekes, J. F., *Am. J. Physiol.,* **127,** 51 (1939).
McCance, R. A., and Wilkinson, E., *J. Physiol.,* **106,** 256 (1947).
Osborn, C. M., and Eversole, W. J., *Federation Proc.,* **8,** 122 (1949).
Osborn, C. M. and LoCascio, L. M., *Anat. Record,* **106,** 62 (1950).
Ragan, C., Ferrebee, J. W., Phyfe, P., Atchley, D. W., and Loeb, R. F., *Am. J. Physiol,* **131,** 73 (1940).
Rice, K., and Richter, C., *Endocrinology,* **33,** 106 (1943).
Robinson, F. J., Power, M. H., and Kepler, E. J., *Proc. Staff Meetings Mayo Clinic,* **16,** 577 (1941).
Roemmelt, J. C., Sartorius, O. W., and Pitts, R. F., *Am. J. Physiol.,* **159,** 124 (1949).
Sanderson, P. H., *Clin. Sci.,* **6,** 197 (1946-48).
Sartorius, O. W., and Pitts, R. F., personal communication, 1949.
Swingle, W. W., Parkins, W. M., Taylor, A. R., and Hays, H. W., *Am. J. Physiol.,* **119,** 557 (1937).
Talbott, J. H., Pecora, L. J., Melville, R. S., and Consolazio, W. A., *J. Clin. Invest.,* **21,** 107 (1942).
Verney, E. B., *Lancet,* **51,** 739 (1946).
Waterhouse, C., and Keutmann, K. H., *J. Clin. Invest.,* **27,** 372 (1948).
White, H. L., Heinbecker, P., and Rolf, D., *Am. J. Physiol.,* **149,** 404 (1947).

DOES METHYL TESTOSTERONE MODIFY THE EFFECTS OF ADRENOCORTICOTROPIC HORMONE (ACTH) AND OF DESOXYCORTICOSTERONE GLUCOSIDE (DOCG)?*

F. C. BARTTER, PAUL FOURMAN, FULLER ALBRIGHT, W. McK. JEFFERIES, ELEANOR DEMPSEY, AND EVELYN L. CARROLL

Harvard Medical School and the Medical Service of the Massachusetts General Hospital, Boston, Massachusetts

The following four premises may be set down:

1. Cushing's syndrome is largely, if not entirely, due to an overproduction of the "carbohydrate-active" steroids ("sugar" hormone) of the adrenal cortex (Albright, 1942-43).
2. The production of the sugar hormone is under the control of the pituitary adrenocorticotropic hormone (ACTH) (Venning, 1948).
3. Therapy with testosterone (Albright *et al.*, 1941) or methyl testosterone (unpublished data) improves certain patients with Cushing's syndrome.
4. Striking objective evidence of this improvement is the nitrogen retention (Albright *et al.*, 1941).

The present study is concerned with the *modus operandi* of testosterone in exerting this beneficial effect. A priori four possibilities come to mind: (*a*) testosterone might block the production of ACTH; (*b*) it might block the production of sugar hormone by ACTH; (*c*) it might block the action of sugar hormone on its end organs; or (*d*) it might simply overcome some of the deleterious effects of too much sugar hormone by an opposite action of its own.

We have sought to eliminate some of these possibilities by comparing the metabolic effects of ACTH given with methyl testosterone with those of ACTH and methyl testosterone given alone. If the action of ACTH is not affected by methyl testosterone, then one might conclude that methyl testosterone does not block the action of ACTH on the adrenals (possibility *b*) or the action of sugar hormone on its end organs (possibility *c*). To explain the action of methyl testosterone one would then have to invoke either an independent opposing action (possibility *d*), a blocking of ACTH production (possibility *a*), or both.

---

* Presented in part before the Association for the Study of Internal Secretions (Bartter *et al.*, 1949).

The expense of these studies was partly defrayed by grants from the Rockefeller Foundation, the National Advisory Cancer Council, Ayerst, McKenna and Harrison, Ltd., and the American Cancer Society on the recommendation of the Committee on Growth of the National Research Council. A bed supported by Mr. Edward Mallinckrodt, Jr., was used for these studies.

The opportunity was taken in this experiment to obtain comparable information about DOCG.

Procedure

The subject of the experiment was a 38-year-old woman with ovarian agenesis, well-trained in the routine of metabolic experiments. For the period of the experiment she was on a constant metabolic regimen (Reifenstein *et al.*, 1945). Each treatment or combination of treatments was given for eight days followed by an eight-day recovery period. There was an initial sixteen-day control period. Urine was collected over two-day periods except for the two days at the beginning of treatment and recovery periods, when it was collected over twenty-four hour periods. It was analyzed for N, P, Ca, K, Na, Cl, Mg, glucose, and 17-ketosteroids. Excretion of reducing steroids (Heard and Sobel, 1946) or "11-oxysteroids" (Talbot *et al.*, 1945) and creatine and creatinine was measured over four-day periods. Feces were collected over four-day periods and analyzed for N, P, Ca, K, and Na. Blood specimens were taken before breakfast on days 1 and 3 of each treatment and recovery period, and the serum was analyzed for P, Ca, glucose, Na, Cl, and Mg. Absolute eosinophil counts were done on alternate days. The sequence of treatments was as follows:

1	Control	8 days
1a	Control	8 days
2	ACTH, 100 mg daily, in 4 doses	8 days
2a	Recovery period	8 days
3	ACTH, 100 mg daily in 4 doses and methyl testosterone, 100 mg daily in 5 doses	8 days
3a	Recovery period	8 days
4	Methyl testosterone, 100 mg daily	8 days
4a	Recovery period	8 days
5	Desoxycorticosterone glucoside, 30 mg daily in 4 doses	8 days
5a	Recovery period	8 days
6	Desoxycorticosterone glucoside 30 mg daily, and methyl testosterone, 100 mg daily	8 days
6a	Recovery period	8 days

With this design (Fisher, 1937) there are three estimates of the effect of methyl testosterone (MT): (1) MT data minus control data, (2) MT with ACTH data minus ACTH data, (3) MT with DOCG data minus DOCG data. We are mainly concerned with the question whether there is any significant difference between (1) and (2).

The design also provides two estimates each of the effects of ACTH and DOCG:

A.	ACTH data minus control data	DOCG data minus control data
B.	ACTH with MT data minus MT data	DOCG with MT data minus MT data

In all these comparisons, there is a possible fallacy in the assumption that successive courses of the same drug have an unchanging effect. How far the assumption is justified, and how the conclusions might be modified where it is not justified, will be discussed.

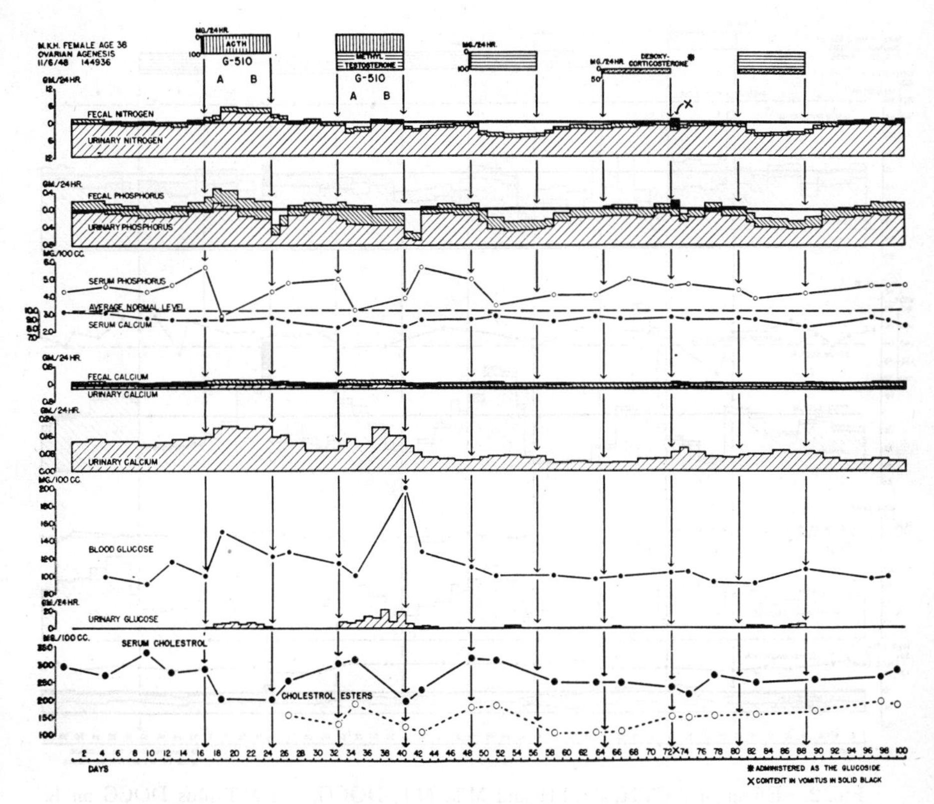

FIG. 1. Effect of ACTH, ACTH plus MT, MT, DOCG, and MT plus DOCG on N, P, and Ca balances, on serum P and Ca, and on blood and urine glucose.

Metabolic data in Figs. 1 and 2 are arranged according to the following scheme:

There is a horizontal base line; intake is charted downward from this base line; the urinary and fecal excretions are then measured upward from the intake line towards the base line. If the output (fecal and urinary) exceeds the intake, the final level will be above the base line; if it does not, the final level will below the base line. Thus a negative balance is indicated by a shaded area above the base line, and a positive balance by a clear area below the base line. The scales for N and P are so chosen that (for changes in protoplasm and bone) the area representing P balance should equal the sum of the corresponding areas for N and Ca; and that representing K balance should equal that for N.

RESULTS

In Figs. 1–3 the data for N, P, Ca, K, Na, and Mg are charted as balance studies, and the remaining data of the experiment are also shown. These charts alone suggest that the effect of ACTH and methyl testosterone given together is the sum of the effects of ACTH and methyl testosterone given separately. Some of the results will be examined in detail by calculating the effects of the different treatments as indicated above. In these calculations, urinary and fecal data have of necessity been taken separately, since there are only two fecal for every five urinary data. The latter have been analyzed statistically (Fourman, *et al.,* see appendix). The known effects of

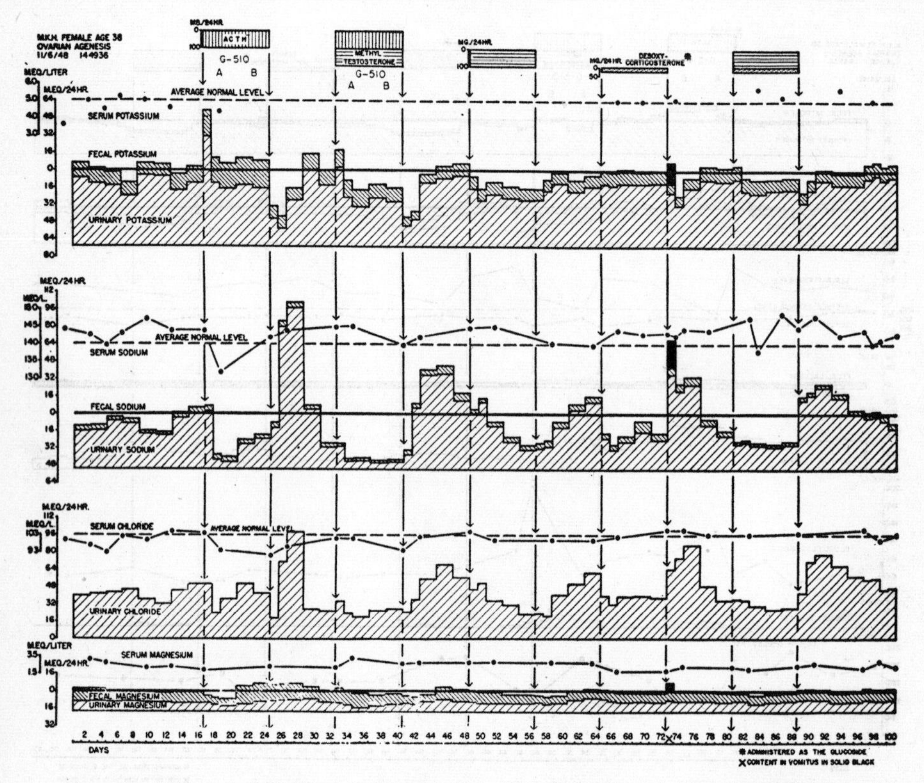

Fig. 2. Effect of ACTH, ACTH and MT, MT, DOCG, and MT plus DOCG on K, Na, and Mg balances, urine chloride, and serum Na, Cl, and Mg. See legend to Fig. 1.

methyl testosterone (Kenyon *et al.,* 1940; Werner and West, 1943; Wilkins *et al.,* 1941) and of ACTH (Bartter *et al.,* 1950; Conn *et al.,* 1948, 1949; Forsham *et al.,* 1948; Mason *et al.,* 1948; Prunty *et al.,* 1948; Venning, 1948) will be considered only briefly.

Effects of Methyl Testosterone (Table 1). The main effect of methyl testosterone was to cause retention of nitrogen and of P, K, and Mg, the constituents of intracellular fluid that were measured, in roughly the proportions in which they occur in protoplasm (Reifenstein *et al.,* 1945) (see Fig. 4). The effect of methyl testosterone given with ACTH was the same (the corresponding data for ACTH having been subtracted) as that of methyl testosterone given alone. The effect of methyl testosterone was significantly less the third time it was given, probably because of a diminishing effect rather than because it was this time given with DOCG.

Another important effect of methyl testosterone was to cause retention of Na and Cl roughly in the proportions in which they occur in extracellular fluid (see Fig. 5).

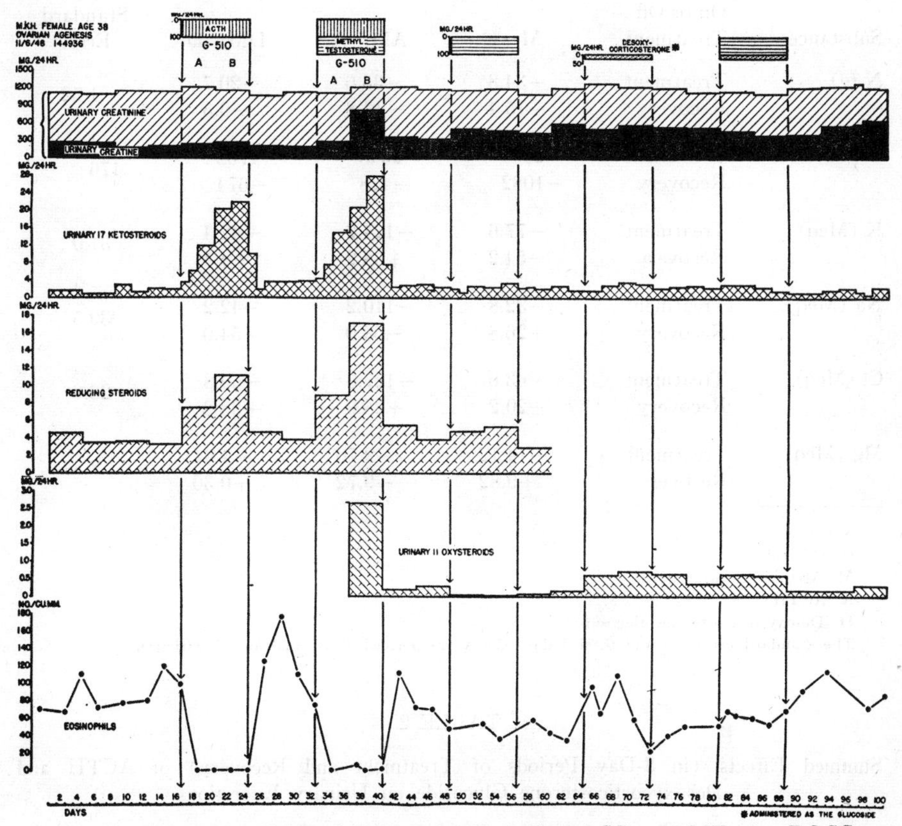

FIG. 3. Effect of ACTH, ACTH and MT, MT, DOCG, and MT plus DOCG on urinary creatine, creatinine, 17-ketosteroids, reducing steroids or "11-oxysteroids," and circulating eosinophils.

Effects of ACTH (Table 2). The effects of ACTH were as previously described (Bartter *et al.*, 1950; Conn *et al.*, 1948, 1949; Forsham *et al.*, 1948; Mason *et al.*, 1948; Prunty *et al.*, 1948; Venning, 1948) and were essentially the same when it was given with methyl testosterone (the corresponding data for methyl testosterone having been subtracted) as when it was given alone (Fig. 6).

The important effects were loss of nitrogen, an early unsustained loss of K, loss of Ca, retention of Na and of proportionately less Cl, raised excretion of steroids, and fall in eosinophils. In addition there was an effect of ACTH not previously described: Mg was retained during treatment and lost during the recovery period. There were not sufficient data for serum Mg (Fig. 2) to say whether or not these changes were due to changes in extracellular fluid content of Mg.

TABLE 1

Summed Effects of Methyl Testosterone on Urinary Excretions in 8-Day Periods of Treatment and Recovery

Substance	On or Off Treatment	M—C	AM—A	DM—D	Standard Error
N (g)	Treatment	−34.8	−35.6	−20.7	1.52
	Recovery	−15.4	−14.8	−0.5	
P (mg)	Treatment	−2594	−2348	−1768	176
	Recovery	−1082	−428	−674	
K (Meq)	Treatment	−77.6	−124.2	−42.4	31.0
	Recovery	−54.2	+38.2	+4.4	
Na (Meq)	Treatment	−32.8	−110.2	−42.2	39.5
	Recovery	+26.8	−68.0	+54.6	
Cl (Meq)	Treatment	−63.6	−127.6	−59.8	26.6
	Recovery	+20.2	+28.0	+38.8	
Mg (Meq)	Treatment	−14.12	−9.70	−4.64	2.84
	Recovery	+9.82	−9.72	+0.30	

Symbols:
C Control
M Methyl testosterone
A ACTH
D Desoxycorticosterone glucoside
The standard errors of the total 8-day effects are derived from analysis of variance.

TABLE 2

Summed Effects (in 8-Day Periods of Treatment and Recovery) of ACTH and Desoxycorticosterone Glucoside on Urinary Excretions

Substance	On or Off Treatment	A—C	AM—M	D—C	DM—M	Standard Error
N (g)	Treatment	24.9	24.0	−10.8	3.3	1.52
	Recovery	3.4	6.1	−4.8	10.2	
P (mg)	Treatment	231	−831	−488	337	176
	Recovery	−503	152	−181	226	
K (Meq)	Treatment	41.9	−4.7	20.5	55.7	31.0
	Recovery	−99.3	−7.1	37.7	20.4	
Na (Meq)	Treatment	−137.5	−213.8	−94.9	−104.5	39.5
	Recovery	289.1	215.3	135.0	162.7	
Cl (Meq)	Treatment	−6.2	−70.3	−36.4	−32.5	26.6
	Recovery	64.9	72.7	151.8	170.3	
Mg (Meq)	Treatment	−19.7	−15.3	−2.0	7.5	2.86
	Recovery	15.6	−4.0	6.6	−2.9	

Effects of DOCG (Table 2). Like ACTH, DOCG caused retention of Na and Cl, and this was not modified significantly by methyl testosterone. In this experiment there were no effects on K and no other effects attributable to DOCG (Fig 6).

M.K.H. FEMALE AGE 38
OVARIAN AGENESIS
NOV. 1948 144936

DIFFERENCES IN
URINARY EXCRETION

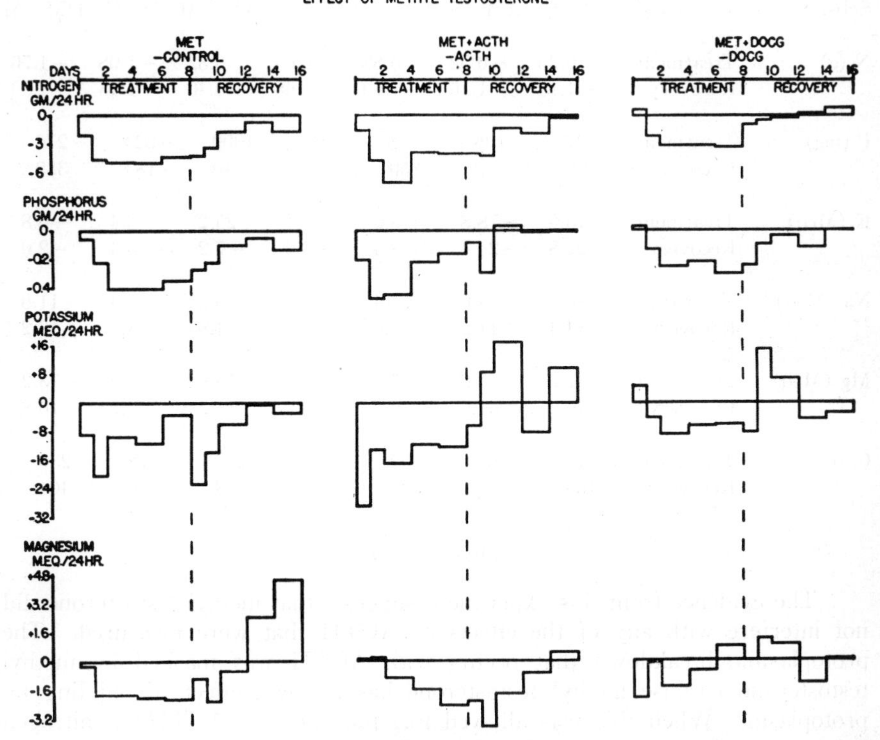

FIG. 4. Effect of methyl testosterone (MET) on urinary N, P, K, and Mg. Retention is shown by negative values, or area below the zero line.

Glycosuria. All three treatments produced glycosuria in varying amounts. With ACTH the glycosuria was accompanied by hyperglycemia, but with all three treatments a lowered renal threshold for glucose was probably a factor in the glycosuria (McAlpine *et al.,* 1948).

Fecal Changes (Table 3). Methyl testosterone had no significant effect on fecal excretions. With ACTH the fecal excretion of P, K, and Ca was higher during treatment periods than in the following recovery periods. We are inclined to think this finding fortuitous, because we have been unable to confirm it in other experiments. The fecal excretion of Mg was high during both ACTH treatment and the following recovery periods. Although these changes could explain some of the fall in urinary Mg excretion during ACTH treatment, they cannot explain the subsequent rise during the recovery period. This suggests a real effect of ACTH on Mg balance.

TABLE 3

Summed Effects (in 8-Day Periods of Treatment and Recovery) of Methyl Testosterone, ACTH and Desoxycorticosterone Glucoside on Fecal Excretions

Substance	On or Off Treatment	M—C	AM—A	DM—D	A—C	AM—M	D—C	DM—M
N (g)	Treatment	−0.25	−0.52	−0.88	0.64	0.32	−1.08	−1.76
	Recovery	−1.44	−1.16	−1.89	−0.68	−0.40	−1.48	1.80
P (mg)	Treatment	−116	−196	−168	1080	1000	−224	−276
	Recovery	−204	−268	350	160	40	−184	380
K (Meq)	Treatment	−15.6	−78.8	−8.0	88.4	25.2	−8.4	−0.8
	Recovery	22.8	−26.4	−8.4	−3.6	−7.2	−16.4	−2.0
Na (Meq)	Treatment	−0.4	−14.0	−24.4	0.4	−13.2	12.4	11.6
	Recovery	14.4	11.2	−17.6	8.0	4.8	18.8	−13.2
Mg (Meq)	Treatment	−7.6	1.2	−7.6	10.8	19.6	13.2	−13.2
	Recovery	−16.0	−31.6	6.0	26.8	11.2	−5.2	16.8
Ca (mg)	Treatment	16	−68	−292	320	236	72	−236
	Recovery	−168	−92	332	−144	−68	−36	464

Discussion

The evidence from this experiment suggests that methyl testosterone did not interfere with any of the effects of ACTH that were measured. The protoplasmic breakdown that occurs with ACTH was masked by methyl testosterone because methyl testosterone has its own effect of building up protoplasm. When this was allowed for, the effect of ACTH on nitrogen balance was the same with and without methyl testosterone. This conclusion was made on the assumption that successive courses of ACTH and of methyl testosterone were directly comparable. In a later study (unpublished) on the same patient this assumption was shown to be warranted for ACTH. In two courses of ACTH (100 mg daily) of eight days each, separated by an eight-day recovery period, the effect on nitrogen balance was the same. But the present experiment suggests that the assumption was not warranted for methyl testosterone, which apparently had a diminishing effect the third time it was given. So it is possible that the data from the second course of methyl testosterone led to an underestimate of its effect in the first course, when it was given with ACTH. If this were so, it would strengthen our conclusion that the effect of ACTH is not diminished by methyl testosterone. Admittedly these arguments carry less weight than would a repetition of the experiment with the treatments given in the reverse order.

The effects of methyl testosterone, ACTH and DOCG on electrolytes may be compared, but with the reservation that for Na and Cl changes the recovery periods were not long enough (Fig. 2).

The changes in N, P, K, and Mg with the different treatments have been charted in Figs. 4 and 6. The scales are in inverse proportion to the concentration of these substances in protoplasm, so that changes only in proto-

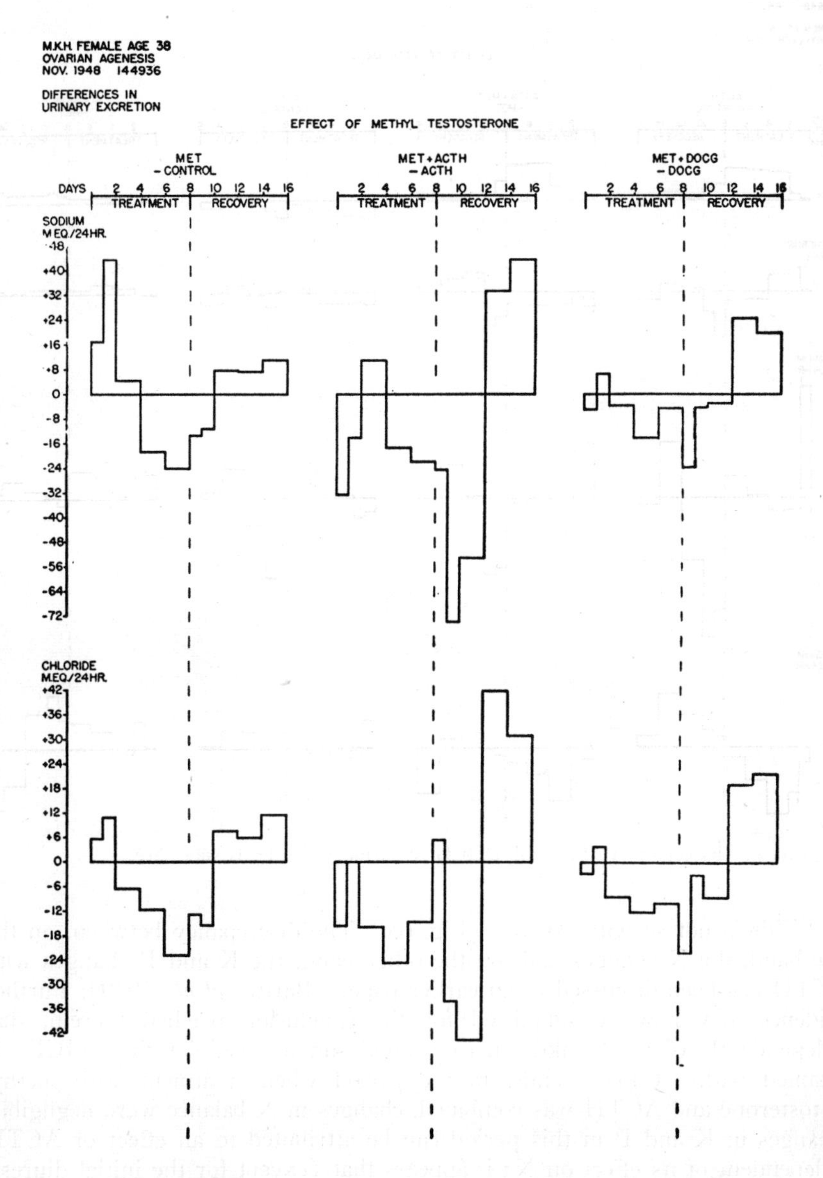

Fig. 5. Effect of methyl testosterone (MET) on urinary Na and Cl.

plasm would cause deviations of similar magnitude for N, P, K, and Mg on the charts (Reifenstein *et al.*, 1945). With methyl testosterone, the changes are those to be expected from changes in protoplasm and even the retention of Na and Cl (Table 1) is of the order expected from the theoretical ECF content of protoplasm (5 ml per gram of nitrogen).

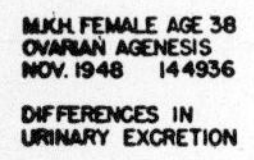

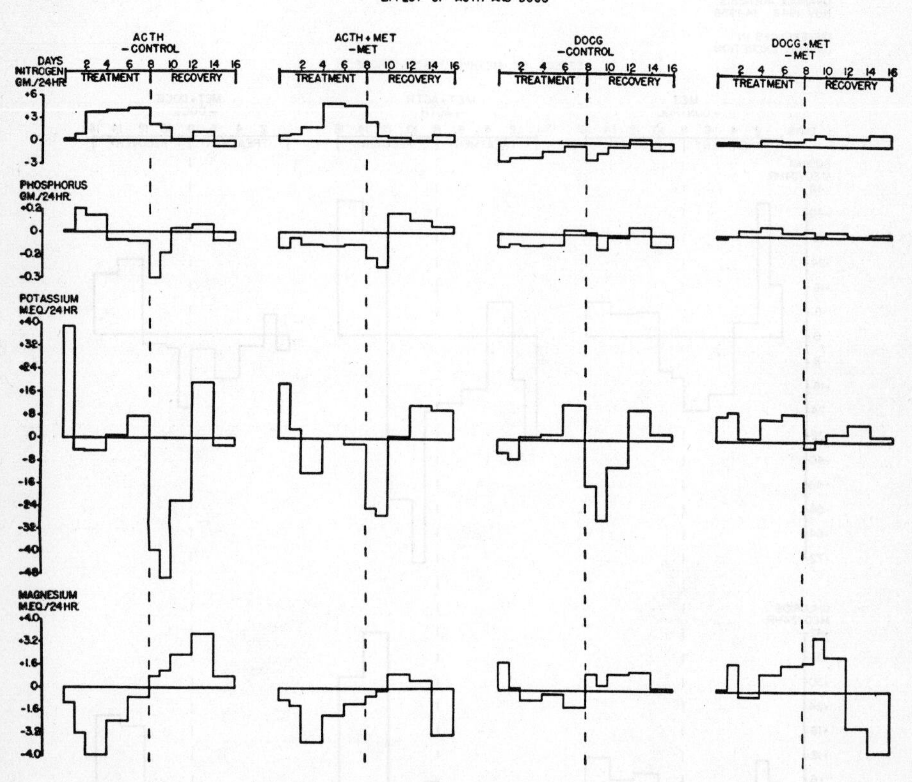

FIG. 6. Effect of ACTH and DOCG on urinary N, P, K, and Mg.

This is not so with ACTH (Fig. 6). The discrepancy between, on the one hand, the N changes and, on the other hand, the K and P changes with ACTH has been discussed in an earlier paper (Bartter *et al.,* 1949). Further evidence may now be submitted for the conclusion reached therein, that independently of the breakdown of protoplasm, intracellular fluid (ICF) is retained with ACTH. Thus, in the period when treatment with methyl testosterone and ACTH was combined, changes in N balance were negligible. Changes in K and P in this period can be attributed to an effect of ACTH independent of its effect on N; it appears that (except for the initial diuresis of K) ACTH caused retention of K and P. These electrolytes are the main cation and anion of ICF, and the effect has been attributed to retention of ICF, presumably with glycogen. The reason the effect of ACTH on K and P does not parallel the effect on N is that it is a resultant of the effect of ACTH on protoplasm (P and K loss) and on glycogen (P and K retention). When the effect of ACTH on protoplasm is masked with methyl testosterone, retention of K and P is demonstrated.

While commensurate changes in K and P probably represent changes in ICF volume, a disproportionate change in K probably represents a change in ICF composition. Thus we interpret the early loss of K with ACTH, with retention on stopping ACTH, as representing a fall and rise in the concentration of K in ICF. When similar changes were produced with 17-hydroxycorticosterone (Fourman *et al.,* 1950), it was shown that the lost K was partly replaced by Na entering the cells. It is likely that in this experiment too Na entered the cells, for the changes in Na greatly exceed those in Cl (Table 2). Mg also presumably entered cells when ACTH was given.

The effect of DOCG on electrolytes differed from that of ACTH. In spite of the fact that retention of Na and Cl was obvious, a loss of K was not demonstrated. Although this difference may be merely a matter of dosage, it stands in contrast to the similarity of the effect on K produced by 17-hydroxycorticosterone and ACTH.

To go back to Cushing's syndrome, these studies suggest that the beneficial action of testosterone is simply to overcome the deleterious effect of sugar hormone on protoplasm by its own effect in the opposite direction. That methyl testosterone does not block the production of sugar hormone is shown by the undiminished rise in steroid excretion. That methyl testosterone does not block the action of sugar hormone on its end organs is shown by the fact that the effect of methyl testosterone on protoplasm was itself offset by ACTH. Since large amounts of ACTH were given, this experiment does not throw light on the question of whether methyl testosterone has a second action—to block the production of ACTH—as suggested by other studies (Bartter, Forbes, Jefferies, Carroll, and Albright, 1949).

The observation that the administration of testosterone with ACTH prevents the loss of protoplasm without interfering with the other known actions of ACTH may have therapeutic implications. For example, with a combination of these two drugs it may be possible to retain the beneficial effects of ACTH on arthritis without producing "Cushing's syndrome medicamentosa."

Summary and Conclusions

1. To help elucidate the beneficial effect of methyl testosterone in Cushing's syndrome, adrenocorticotropic hormone (ACTH) and methyl testosterone were given alone and in combination to one subject. The daily dose of each drug was 100 mg. Each course of treatment lasted for eight days and was followed by an eight-day recovery period. The opportunity was taken to obtain comparable information about desoxycorticosterone glucoside.

2. On a constant intake, the fecal and urinary excretion of N, P, Ca, K, Na, and Mg, the urinary excretion of Cl, glucose, 17-ketosteroids and reducing

steroids, the blood glucose and serum P, Ca, Na, and Cl levels were measured. Eosinophil counts were done.

3. With ACTH there were:
   (*a*) A loss of protoplasm.
   (*b*) A relative retention of intracellular fluid (with glycogen?).
   (*c*) A relative loss of K from intracellular fluid.
   (*d*) A retention of Na and Mg, presumably in intracellular fluid.

4. With methyl testosterone there was simply a retention of protoplasm.

5. With ACTH and methyl testosterone together the change in protoplasm was a summation of their separate opposite effects. None of the other effects of ACTH was changed by methyl testosterone.

6. Testosterone exerts its beneficial effect in Cushing's syndrome by offsetting (not preventing) the deleterious effect of sugar hormone on protoplasm. Whether in addition it inhibits the production of ACTH these studies do not show.

REFERENCES

ALBRIGHT, FULLER, Cushing's syndrome. Harvey Lectures, 1942-43, **Ser. 38,** 123.

ALBRIGHT, F., PARSON, W., BLOOMBERG, E., Therapy in Cushing's snydrome. *J. Clin. Endocrinol.,* **1,** 375 (1941).

BARTTER, F. C., FORBES, A. P., JEFFERIES, W. M., CARROLL, E. L., ALBRIGHT, F., The mechanism of action of testosterone in the therapy of Cushing's syndrome. *J. Clin. Endocrinol.,* **9,** 663 (1949).

BARTTER, F. C., FOURMAN, P., FORBES, A. P., JEFFERIES, W. M., ALBRIGHT, F., An Analysis of the Various Factors Influencing Potassium, Phosphorus, and Sodium Excretion on Administration of ACTH. First Conference on Metabolic Interrelations, p. 137. Josiah Macy Jr. Foundation, New York, 1949.

BARTTER, F. C., FOURMAN, P., ALBRIGHT, F., FORBES, A. P., JEFFERIES, W. M., GRISWOLD, G., DEMPSEY, E., BRYANT, D., and CARROLL, E., The effect of adrenocorticotropic hormone in panhypopituitarism. *J. Clin. Invest.,* **29,** 1950 (1950).

CONN, J. W., LOUIS, L. H., and WHEELER, C. E., Production of temporary diabetes mellitus in man with pituitary adrenocorticotropic hormone; relation to uric acid metabolism. *J. Lab. Clin. Med.,* **33,** 1948 (651).

CONN, J., LOUIS, L. H., and JOHNSTON, M. W., Metabolism of uric acid, glutathione, and nitrogen, and excretion of 11-oxysteroids and 17-ketosteroids during induction of diabetes mellitus in man with adrenocorticotropic hormone. *J. Lab. Clin. Med.,* **34,** 1949 (255).

FISHER, R. A., *The Design of Experiments.* Second Edition. Oliver and Boyd, Edinburgh, 1937.

FORSHAM, P. H., THORN, G. W., PRUNTY, F. T. G., and HILLS, A. G., Clinical studies with pituitary adrenocorticotropin, *J. Clin. Endocrinol.,* **8,** 1 (1948).

FOURMAN, P., BARTTER, F. C., ALBRIGHT, F., DEMPSEY, E., CARROLL, E., ALEXANDER, J., Effects of 17-hydroxy-corticosterone ("Compound F") in man. *J. Clin. Invest.,* 1950. In press.

HEARD, R. D. H., and SOBEL, H., A colorimetric method for the estimation of reducing steroids. *J. Biol. Chem.,* **165,** 687 (1946).

KENYON, A. T., KNOWLTON, K., SANDIFORD, I., KOCH, F. C., and LOTWIN, G., a comparative study of the metabolic effects of testosterone propionate in normal men and women and in eunuchoidism. *Endocrinology,* **26,** 26 (1940).

MASON, H. L., POWER, M. H., RYNEARSON, E. H., CIARARMELLI, L. C., LI, C. H., and EVANS, H. M., Results of administration of anterior pituitary adrenocorticotropic hormone to a normal human subject. *J. Clin. Endocrinol.,* **1**, 1, (1948).

MATHER, K., Statistical Analysis in Biology. Interscience Publishers, Inc., New York, 1947.

MCALPINE, T. H., VENNING, E. H., JOHNSON, L., SCHENKER, V., HOFFMAN, M. M., and BROWNE, J. S. L., Metabolic changes following the administration of pituitary adrenocorticotropic hormone (ACTH) to normal humans. *J. C. E.,* **8**, 591 (1948).

PRUNTY, F. T. G., FORSHAM, P. H., and THORN, G. W., Desoxycorticosterone-like activity induced by adrenocorticotrophin in man, *Clin. Sci.,* **7**, 109 (1948).

REIFENSTEIN, E. C., ALBRIGHT, FULLER, WELLS, S. L., The accumulation, interpretation, and presentation of data pertaining to metabolic balances, notably those of calcium, phosphorus, and nitrogen. *J. Clin. Endocrinol.,* **5**; 367 (1945).

TALBOT, N. B., SALTZMAN, A. H., WIXOM, R. L., WOLFE, J. K., The colorimetric assay of urinary corticosteroid-like substances. *J. Biol. Chem.,* **160**, 535 (1945).

VENNING, E. H., The Effect of Large Doses of Adrenocorticotrophin on Steroid Metabolism. Seventeenth Conference on Metabolic Aspects of Convalescence, Josiah Macy Jr. Foundation, New York, 1948.

WERNER, S. C., and WEST, R., Nitrogen retention, creatinuria, and other effects of the treatment of Simmonds' disease with methyl testosterone. *J. C. I.,* **22**, 335 (1943).

WILKINS, L., FLEISCHMAN, N. W., and HOWARD, J. E., Creatinuria induced by methyl testosterone in the treatment of dwarfed boys and girls. *Bull. Johns Hopkins Hosp.,* **69**, 493 (1941).

EVALUATION OF THE METABOLIC EFFECTS OF A PEPTIDE MIXTURE DERIVED FROM PURE ADRENOCORTICOTROPIC HORMONE*

LAURANCE W. KINSELL, CHOH HAO LI, SHELDON MARGEN, GEORGE D. MICHAELS, AND ROBERT N. HEDGES

WITH THE TECHNICAL ASSISTANCE OF

CARL T. ANDERSON, MAXINE E. HUTCHIN, AND JUDITH LANGE

University of California-U. S. Naval Hospital, Oakland, and the University of California, Berkeley

In 1948, Li reported that a peptide mixture, prepared by hydrolyzing whole adrenocorticotropic hormone, retained its ability to stimulate the adrenal cortices in hypophysectomized rats (Li, 1948). In March, 1949 (Li, 1949), he further reported some of the physical characteristics of this peptide mixture. The present report represents a study in a 60-year-old male patient with classical rheumatoid arthritis who, after preliminary equilibration upon a chemically constant food intake, received 100 mg of such peptide mixture daily for 5 days. After an interval of 15 days, during which time the patient continued under the balance study routine, he received 100 mg of whole pituitary adrenocorticotropin daily for an additional 5-day period.

In the human subject, adrenocorticotropin preparations have been shown to increase the urinary excretion of 17-ketosteroids, to cause significant sodium and water retention, to cause an increased excretion of urinary nitrogen, potassium, calcium, and phosphorus, to cause significant elevation in fasting blood sugar and a striking fall in circulating eosinophils.

In Figs. 1-3 are shown the changes in these entities in the patient above noted during the administration of the peptide mixture and of whole adrenocorticotropic hormone. The following changes were noted in response to ACTH peptide administration:

1. Elevation in urinary 17-ketosteroids from values averaging less than 3 to a level of 36 mg per 24 hours.
2. A profound decrease of urinary sodium excretion coupled with a proportional retention of water and gain in body weight.
3. Considerable elevation of fasting blood sugar (Fig. 4).
4. A fall in circulating eosinophils to zero (Fig. 4).
5. Slight but significant increase in excretion of calcium and magnesium. The increase in calcium excretion was more marked in the stool than in the urine.

* This work is supported by grants from the Research Division of the Bureau of Medicine and Surgery, U. S. Navy (BuMed #007046), from the Office of Naval Research, under a contract between the latter and the University of California, and from the U. S. Public Health Service Contract No. RG-409, Research Board of the University of California, Berkeley, California.

Some of this material has been published in *Proceedings of the First ACTH Conference,* Mote. Published by Blakiston, 1950.

It will be noted that much the same qualitative changes occurred during the administration of whole adrenocorticotropin. The increase in 17-ketosteroid excretion was, however, of greater magnitude, and the sodium retention was much less in proportion to the retention of water and gain in body

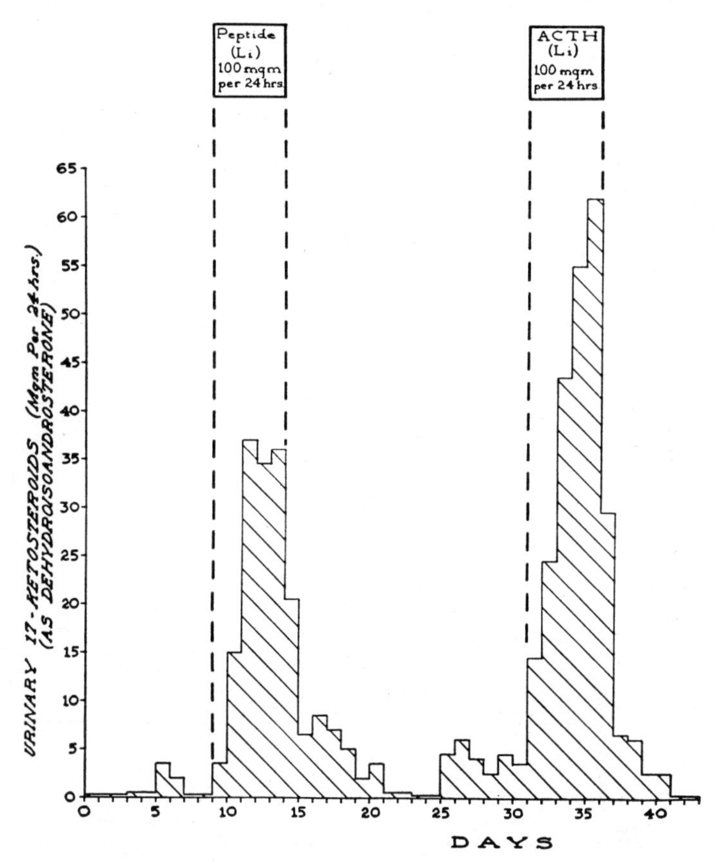

FIG. 1. Urinary 17-ketosteroid excretion in response to ACTH peptide and to whole pituitary adrenocorticotropic hormone.

weight. There was a significant increase in potassium excretion, and the elevation of fasting blood sugar was greater than that which occurred during peptide administration. In Fig. 3 are shown the changes in nitrogen balance and in phosphorus balance during the administration of the ACTH peptide and of the whole ACTH. It will be noted that during the administration of both substances a significant negativity of nitrogen and of phosphorus balances was observed.

It was further found that this patient had a considerable increase in urinary excretion of both organic and inorganic sulfur during the administration of the peptide and of the whole ACTH. The increase in organic sulfur was found to be partially attributable to increased cystine and methionine excretion. Considerably more than 50% of the total increase in urinary organic sulfur could not be accounted for by these two amino acids. The chemical nature of the remaining material is at the present time unknown.

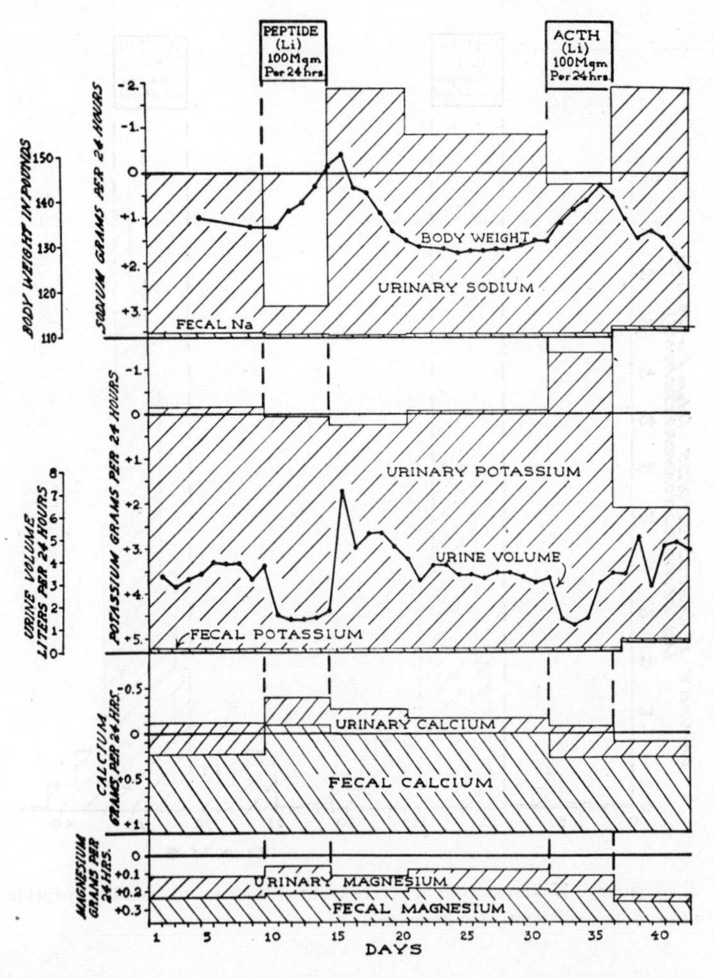

Fig. 2. Changes in sodium, potassium, calcium, and magnesium balance in response to ACTH peptide and whole ACTH (these and following balance charts are so constructed that a clear area below the 0 line represents positive balance and a hatched area above the 0 line represents negative balance).

It should also be noted that during the administration of the peptide mixture this patient's arthritic manifestations improved rapidly. When the peptide was discontinued the joints again became red, swollen, and painful. Administration of whole adrenocorticotropin again resulted in a rapid improvement of the inflammatory process in the joints.

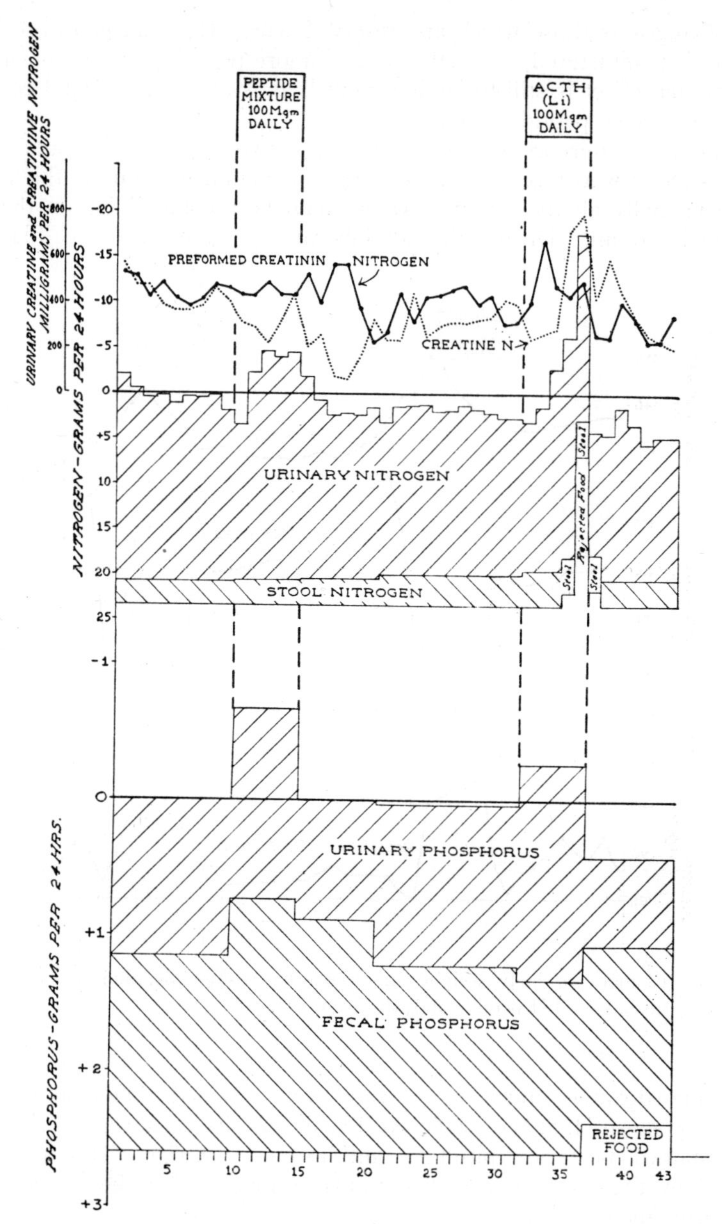

FIG. 3. Changes in nitrogen and phosphorus balance and in creatine and creatinine excretion in response to ACTH peptide and whole ACTH.

DISCUSSION

From the findings above reported it is apparent that a mixture of peptide fragments derived from whole adrenocorticotropin, the fragments having an average molecular weight of 1200, results in much the same metabolic and clinical changes as those which are observed when whole adrenocorticotropic hormone is administered. Whether one or more peptides are responsible for this hormonal effect is still to be determined, as is also the size and composition of the specific peptide material.

A notable difference in the effect of the two materials is found in the relative sodium-water retention. During the administration of the peptide mixture virtually all the retained water could be accounted for on the basis of retained sodium. During the administration of the whole ACTH, however, proportionately much more water was retained than sodium. This presumably means that the whole ACTH despite its "electrophoretic purity" still contained posterior pituitary antidiuretic material. This concept is bolstered by the occurrence of abdominal cramps during the administration of the latter material.

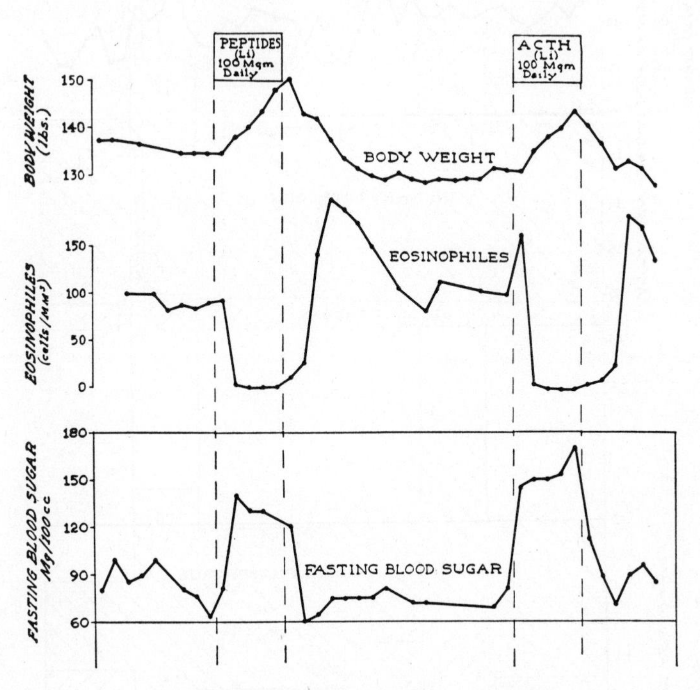

FIG. 4. Changes in body weight, in circulating eosinophils, and in fasting blood sugar in response to ACTH peptide and whole ACTH.

Adrenocorticotropic hormone derived from natural sources would never be available in sufficient quantities to supply the clinical requirements for this material. The demonstration that peptide fragments derived from whole adrenocorticotropin retain a high degree of hormonal activity suggests the possible synthesis of such material. It seems probable that this will eventually be effected. Needless to say, the amount of time necessary to achieve such a synthesis is still completely unknown.

Summary

A protein-free peptide mixture prepared by enzymatic hydrolysis of whole adrenocorticotropin has been found to possess a high degree of adrenocorticotropic activity when administered to a human subject suffering from rheumatoid arthritis. The metabolic and clinical changes observed were comparable to those which occurred during the administration of whole adrenocorticotropin.

REFERENCES

Li, C. H., Conference on Metabolic Aspects of Convalescence. Josiah Macy Jr. Foundation, **17**, 114, March, 1948.

Li, C. H., Relative size of adrenocorticotrophically active peptide fragments, *Federation Proc.*, **8**, 1949.

OBSERVATIONS ON THE METABOLIC EFFECTS OF CORTISONE AND ACTH IN MAN

RANDALL G. SPRAGUE AND MARSCHELLE H. POWER

Mayo Clinic, Rochester, Minnesota

Our purpose is to present some representative data from recent observations of the metabolic effects of ACTH* (pituitary adrenocorticotropic hormone) and cortisone.† Virtually all the observations to be reported were made on patients with rheumatoid arthritis. The collaboration of Drs. P. S. Hench, C. H. Slocumb, and H. F. Polley in these investigations is gratefully acknowledged. The metabolic changes to be described are principally those associated with prolonged (in contrast to short-term) administration of the hormones.

Balance studies for potassium, sodium, and chloride were carried out, and the concentrations of the blood electrolytes were determined at intervals. Balances for nitrogen, calcium, and phosphorus also were measured. Urinary creatine, creatinine, and uric acid were determined. In many of the cases blood chemical determinations for albumin, globulin, calcium, phosphorus, and uric acid were performed. The blood sugar was estimated periodically in most cases. In 3 cases the glutathione in erythrocytes and whole blood was determined at intervals. Zinc sulfate turbidity of serum, an approximate measure of gamma globulins, was determined in several cases. References to the chemical methods employed are omitted here in the interests of brevity, but are presented in another publication (Sprague and associates, 1950).

Observations were made on 33 patients (cases 1 to 33). In this paper case numbers are the same as those used in the paper just mentioned (Sprague and associates, 1950).

Balance Studies

Cortisone‡ was administered intramuscularly, for purposes of metabolic study,§ in doses of 100 and 200 mg daily, in two doses as a saline suspension

* The authors are indebted to Armour and Company and to Dr. John R. Mote, Director of Medical Research, for the supplies of ACTH used in this study.

† The authors are indebted to Merck & Co., Inc. and Dr. James M. Carlisle, Medical Director, for supplies of cortisone.

‡ Both cortisone and cortisone acetate were employed in different studies. Inasmuch as there seem to be no important differences in the physiologic effects of the two compounds, both will be referred to as cortisone. It is realized that 100 mg of cortisone acetate presumably has the same physiologic activity as 89 mg of cortisone, and that there may be minor differences in the rate of absorption of the two compounds from sites of injection due to different solubilities.

§ The authors are indebted to Miss Gordon Sampson, chief dietitian, Mayo Clinic metabolism unit, St. Marys Hospital, for her careful planning and preparation of the diets used in metabolic balance studies.

of finely ground crystals containing 25 mg per cubic centimeter. ACTH* was administered intramuscularly in three or four doses totaling 100 or 105 mg a day. One patient received estrone simultaneously with cortisone on one occasion, and testosterone propionate with cortisone on another. Another patient received 900 mg of 17-hydroxycorticosterone (compound F)† in twelve days.

In the cases of 2 men with rheumatoid arthritis (cases 4 and 5), it was found that the metabolic effects of 100 mg of cortisone daily for thirty and twelve days, respectively, as far as they were measured, were not pronounced. Mild alkalosis developed in case 4. Nevertheless, striking improvement of the rheumatoid arthritis occurred in both cases, as described by Hench, Kendall, Slocumb, and Polley (1949).

More detailed data will be presented in 3 cases in which the patients received ACTH and larger doses of cortisone, resulting in more significant metabolic alterations.

Case 14. The patient was a woman, 45 years of age, with rheumatoid arthritis of five years' duration. Four years before our studies menopause had been induced by roentgen therapy of the ovaries.

In this case changes in the plasma electrolytes and *p*H of the blood during and following administration of several hormones were pronounced. During the administration of 100 mg of ACTH daily for twelve days, the concentration of plasma bicarbonate increased, that of potassium and sodium decreased and that of chloride decreased (Fig. 1). The most marked changes were observed at the end of the period of administration of ACTH. Six days after cessation of administration of the hormone, normal concentrations of plasma electrolytes were found.

The balances for potassium, sodium, and chloride, as well as alterations in blood pressure and body weight during the development of the foregoing changes in the plasma electrolytes, are charted in Fig. 2.‡ Marked retention of sodium and chloride and slight loss of potassium occurred on the first two days of administration of ACTH. Following this, there was an increase in the excretion of sodium, potassium, and chloride above the control levels, resulting in markedly negative balances of the latter two ions during the

* Different preparations of ACTH contained from less than 0.4 to 3.3 units of oxytocin per vial. Each vial contained the equivalent of 25 mg of Armour standard ACTH LA-1-A, except the vials used in case 11, which contained 35 mg. The pressor content varied from less than 0.5 unit to 5 units per vial.

† The authors are indebted to the Upjohn Company and Dr. E. Gifford Upjohn for the 17-hydroxycorticosterone (compound F) which was used in the study of this patient.

‡ In this and subsequent charts of balance data, daily intake is charted from the 0 line downward and average daily excretion (feces below, urine above) from the bottom line upward. A negative balance is, therefore, indicated by extension of the column above the 0 line, and a positive balance by a clear area below the 0 line. Because the urinary excretion of electrolytes (particularly sodium and chloride) in some instances varied widely from day to day within a single six-day period, the figures for daily urinary excretion of potassium, sodium, and chloride are shown in some periods, and in others the mean values for three days.

twelve days of administration of ACTH. When administration of ACTH was stopped, marked retention of chloride and potassium and slight retention of sodium, occurred, adequate to compensate for the previous loss of these electrolytes.

Subsequently, cortisone was administered in a dose of 100 mg daily for eighteen days followed immediately by 200 mg daily for twelve days. Slight changes in the plasma electrolytes occurred during eighteen days when the

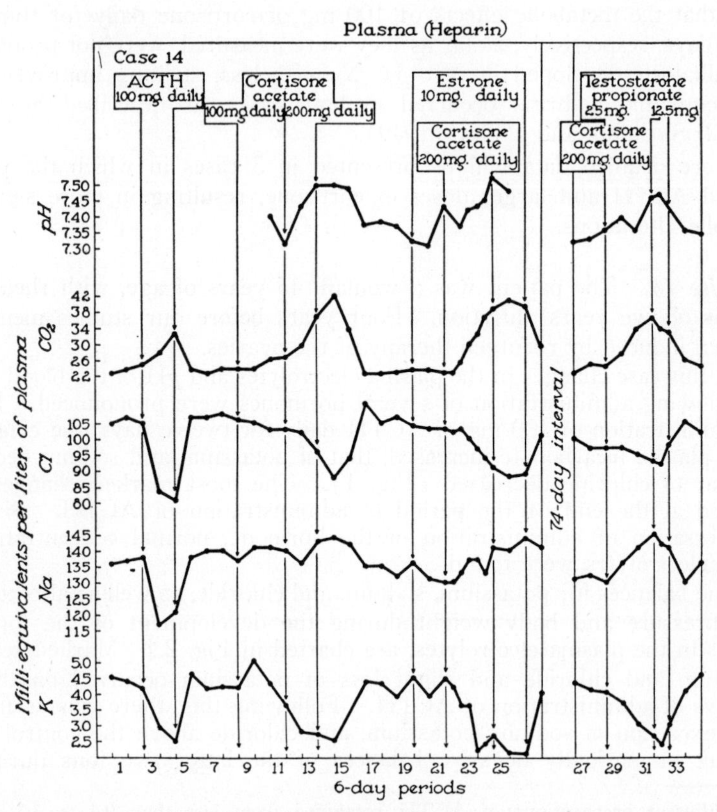

FIG. 1 (case 14). Alterations in plasma electrolytes by ACTH, cortisone, cortisone plus estrone, and cortisone plus testosterone propionate.

dose was 100 mg daily (Fig. 1). When the dose was increased to 200 mg daily, however, a rather severe hypochloremic, hypokaliemic alkalosis developed, similar in all respects to the alkalosis which is observed in some cases of Cushing's syndrome. The most marked changes were observed twelve days after administration of the hormone was stopped, presumably because of prolonged absorption from sites of injection. Then the values gradually reverted to normal. There was a slight increase in plasma sodium following administration of cortisone, in contrast to the decrease with ACTH. There was, as a consequence of this, a more marked rise in plasma bicarbonate under the influence of cortisone than had previously been observed under the influence of ACTH. Negative balances for potassium, chloride, and sodium eventually developed (Fig. 2). The loss of sodium did not become apparent until the first six-day period after administration of the hormone was stopped. The excretion of chloride, on the other hand, increased slightly beyond the intake in the last period of administration of cortisone (that is, period 13). In

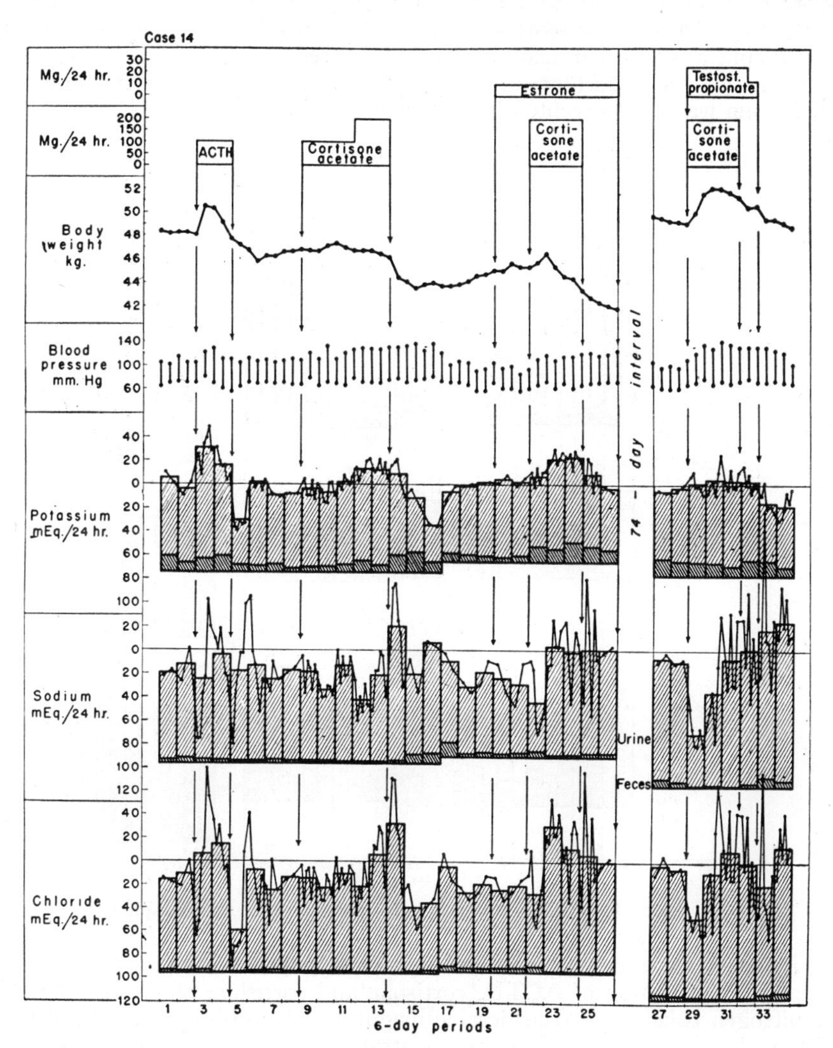

FIG. 2 (case 14). Body weight, blood pressure, and balance data for potassium, sodium, and chloride.

the first period after cessation of the administration of the hormone (that is, period 14), the chloride balance became more markedly negative. It is thought that the loss of sodium and chloride in period 14 was probably due to a continuing hormonal effect rather than to withdrawal of the hormone. In period 15, when the physiologic effect of the hormone was presumably waning, there was retention of sodium and chloride.

Simultaneous administration of estrone in a dose of 10 mg daily did not modify the effects of 200 mg daily of cortisone on the concentrations of the plasma electrolytes and on electrolyte balances (Figs. 1 and 2).

Simultaneous administration of 200 mg of cortisone and 25 mg of testosterone propionate obviated the loss of nitrogen which was previously observed during administration of cortisone alone (see next paragraph). The loss of potassium was less than had previously occurred with cortisone alone. The development of a hypochloremic, hypokaliemic alkalosis under the influence of cortisone was not prevented by testosterone (Fig. 1).

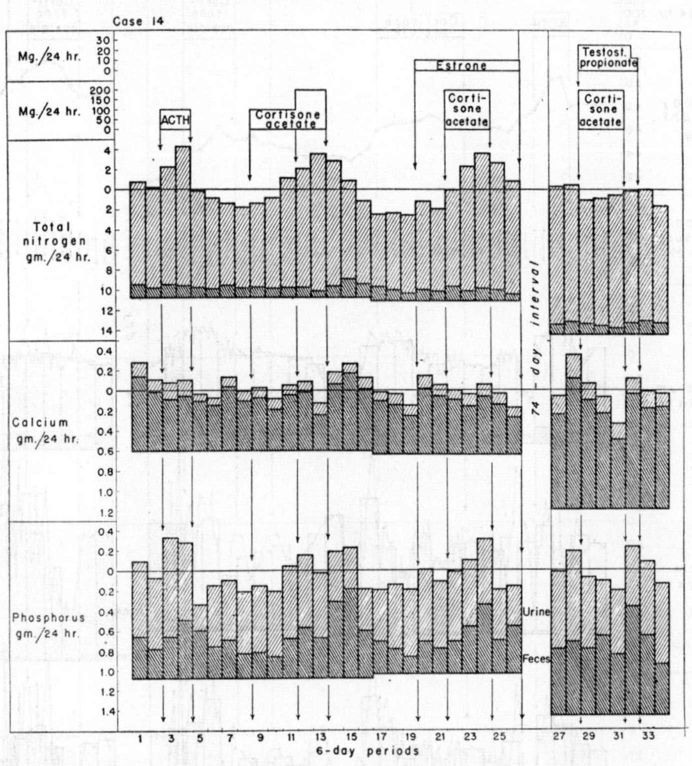

FIG. 3 (case 14). Balance data for nitrogen, calcium, and phosphorus.

Balances for nitrogen, calcium, and phosphorus are charted in Fig. 3. During administration of ACTH, cortisone and cortisone plus estrone, negative nitrogen balances of considerable magnitude were observed. As already mentioned, the simultaneous administration of testosterone propionate prevented an increase of excretion of nitrogen. Increases in urinary calcium and phosphorus were generally slight or absent during administration of cortisone. During administration of ACTH, significant increases in urinary calcium and phosphorus occurred. Significant increases in fecal phosphorus were observed during and after all four programs of treatment.

Case 6. The patient was a woman, 49 years of age, with chronic rheumatoid arthritis of three years' duration. Studies were made of the metabolic effects of ACTH, 17-hydroxycorticosterone (compound F) and cortisone. Alterations in the concentrations of the plasma electrolytes during and after administration of ACTH in a dose of 100 mg daily were qualitatively similar to those previously observed in cases 11 and 14; plasma sodium decreased, chloride decreased to a greater degree, bicarbonate increased moderately and potassium decreased. Thus, a hypochloremic, hypokaliemic alkalosis developed which was similar to that which is observed in some cases of Cushing's syndrome except that the concentration of plasma sodium decreased.

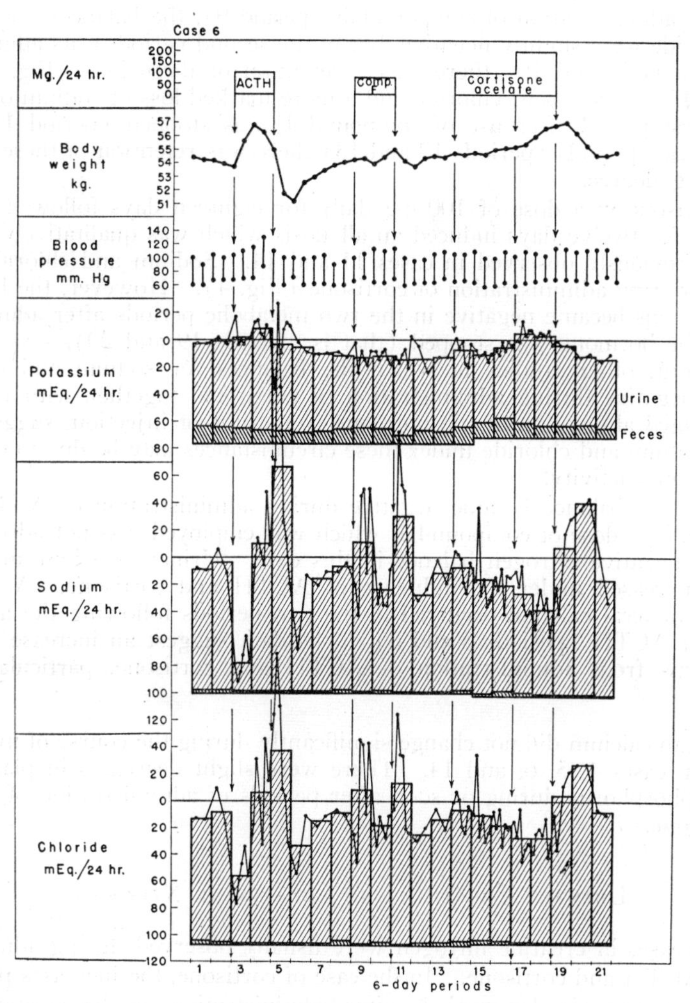

FIG. 4 (case 6). Body weight, blood pressure, and balances of potassium, sodium, and chloride.

The balance data (Fig. 4) indicate that in the first period of administration of ACTH (period 3) there was marked retention of sodium and chloride but that in the second period (period 4), the balances for these ions became negative. Marked loss of chloride and an even more marked loss of sodium occurred on cessation of administration of ACTH (period 5). In period

6, retention of sodium and chloride again occurred. ACTH induced, during the two periods of its administration, a moderate loss of potassium. This was followed in succeeding periods by retention of potassium.

Administration of compound F in an average daily dose of 75 mg for twelve days resulted in the development of an alkalosis of mild degree attributable to slight lowering of the level of the plasma chlorides. There was no significant change in the level of plasma sodium or potassium. In the first period of administration of compound F (period 9), the balances for sodium and chloride were slightly negative, but in the second period of its administration (that is, period 10) there was a retention of these ions (Fig. 4). A "rebound" with a loss of chloride and a more marked loss of sodium occurred in the first period after use of compound F was stopped (period 11). In the next two periods (periods 12 and 13) there was retention of these ions of decreasing degree.

Cortisone in a dose of 100 mg daily for eighteen days followed by 200 mg daily for twelve days induced an alkalosis which was qualitatively similar to that previously observed in cases 11 and 14. Sodium and chloride were retained during administration of cortisone (Fig. 4). However, the balances for these ions became negative in the two metabolic periods after administration of the hormone was stopped (that is, periods 19 and 20). As already pointed out, other studies in which the balances for sodium and chloride became negative during administration of cortisone, together with evidence of prolonged absorption of the hormone from sites of injection, suggest that loss of sodium and chloride under these circumstances may be due to continuing hormonal activity.

Nitrogen balance became negative during administration of ACTH and cortisone. The dose of compound F which was employed was not adequate to induce a negative nitrogen balance in this case. Urinary calcium and phosphorus increased under the influence of ACTH and cortisone. A marked decrease of fecal calcium occurred in the two periods following the administration of ACTH (periods 5 and 6). The data suggest an increase in fecal phosphorus from administration of ACTH and cortisone, particularly the latter.

Serum calcium did not change significantly during the course of metabolic studies in cases 4, 5, 6, and 14. There were slight decreases in plasma inorganic phosphorus during or soon after periods of administration of ACTH or cortisone.

Urinary Creatine and Creatinine Nitrogen

Increases in creatine nitrogen were usually observed during administration of ACTH and cortisone. In the case of cortisone, the increases persisted for one or two six-day periods after administration of the hormone was stopped. Representative data obtained in case 14 are charted in Fig. 5. In this case an increase in creatine nitrogen during and after administration of cortisone was not prevented by prior and simultaneous administration of estrone. Simultaneous administration of testosterone propionate, on the other hand, initially caused a decrease in creatine nitrogen, and later apparently prevented in part the rise which customarily followed administration of cortisone.

With the possible exception of case 14, there were no changes in urinary preformed creatinine nitrogen which could be attributed with certainty to ACTH or cortisone. In case 14, during administration of cortisone plus testosterone propionate, there was a significant rise in preformed creatinine nitrogen. The previous absence of significant increase in preformed creatinine nitrogen during administration of ACTH or cortisone alone suggests that testosterone rather than cortisone was responsible.

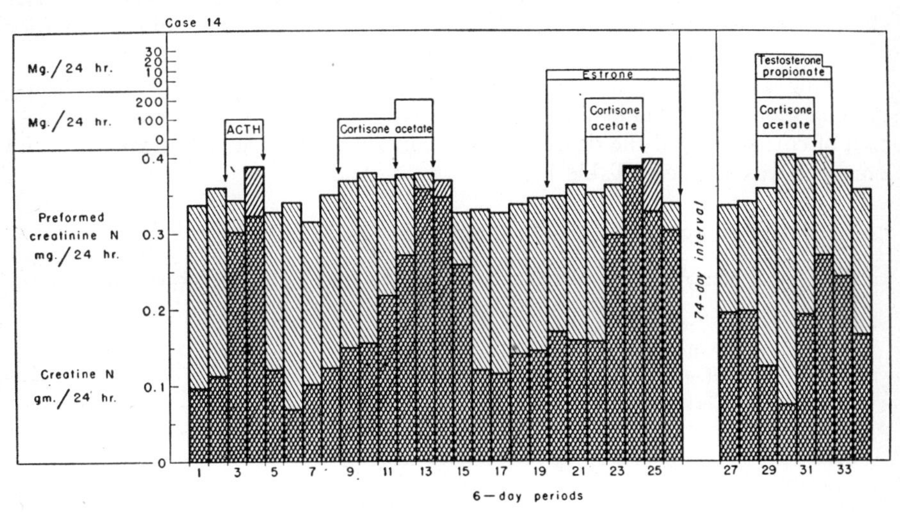

FIG. 5 (case 14). Effect of ACTH, cortisone, cortisone plus estrone, and cortisone plus testosterone propionate on urinary creatine and preformed creatinine, expressed as nitrogen. The double cross-hatched area represents creatine and the total height of each column represents performed creatinine nitrogen except in period 4, when creatine exceeded preformed creatinine.

URIC ACID IN SERUM AND URINE

Excretion of uric acid was studied in 5 cases. In general, when the dose of cortisone was 100 mg daily there were slight increases in urinary uric acid; when the dose was 200 mg daily, or when ACTH in a dose of 100 or 105 mg daily was administered, there were usually more marked increases. Decreases in serum uric acid of significant degree occurred with both ACTH and cortisone. However, in other cases increases in urinary uric acid were observed unassociated with decreases in serum uric acid, particularly if the serum level was initially within the normal range.

SERUM ALBUMIN, GLOBULIN, AND ZINC SULFATE TURBIDITY

During administration of cortisone, as previously noted by Hench and his associates (1949), elevated values of serum globulin and depressed levels of serum albumin usually changed toward normal. In Table 1 is summarized a considerable mass of data on serum albumin and globulin in all the cases in which they were studied. It will be noted that the mean difference in the values of serum albumin obtained during administration of cortisone, as contrasted to the values obtained prior to administration of the hormone, was slightly positive. The increase was maintained when administration of the hormone was discontinued. The mean differences in the values of serum globulin were greater and in the opposite direction. In other words, the effect of cortisone was to increase serum albumin slightly, and to decrease serum globulin to a greater degree. Examination of the data in individual cases revealed that the aforementioned changes were most marked when the pretreatment values for serum albumin were depressed and those for serum globulin elevated.

Zinc sulfate turbidity, which presumably is a rough measure of serum gamma globulins (Kunkel, 1947), decreased significantly during administration of cortisone, and the decrease was maintained after administration of the hormone was stopped (Table 1).

TABLE 1

Changes in Serum Albumin, Globulin, and Zinc Sulfate Turbidity during Administration and after Withdrawal of Cortisone

	Mean Value before Cortisone	During Cortisone		After Cortisone	
		Cases	Mean differences*	Cases	Mean differences*
Albumin	4.0†	26	+0.18 ±0.07†	28	+0.22 ±0.9†
Globulin	3.1†	26	−0.67 ±0.08†	28	−0.72 ±0.10†
Zinc sulfate turbidity	20‡	20	−7.0 ±1.8‡	16	−9.1 ±2.8‡

Note. The available data for serum albumin, globulin, and zinc sulfate turbidity were treated as follows: The observed values before administration of cortisone was instituted, during the time that the hormone was being administered and after administration was stopped, were contrasted. The pertinent differences were calculated for each individual, that is, the difference between the values obtained before and during administration, and the difference between the values obtained before and after administration. The mean of these differences, together with its standard error, was then calculated from the individual differences. The authors wish to acknowledge the assistance of the Division of Biometry and Medical Statistics in making the statistical calculations.

* A plus sign (+) indicates an increase as compared with the level before administration of cortisone, a minus sign (—) indicates a decrease. The value following the ± sign is the standard error of the mean difference. The mean difference is considered statistically significant if it is twice its standard error. Therefore, all of the changes tabulated above are considered statistically significant.

† Grams per 100 cc of serum.

‡ Units.

Blood Glutathione

Determinations of glutathione in whole blood and erythrocytes were made in 3 cases at intervals of six days throughout the course of metabolic studies. In each case there was a gradual rise in the concentration of glutathione in erythrocytes associated with a progressive drop in concentration of hemoglobin, erythrocyte count and hematocrit reading. There was little change in the concentration of glutathione in whole blood. During administration of hormones, there was no consistent pattern of change in the 3 cases. For example, in 2 cases during administration of ACTH, there was a slight increase in glutathione in whole blood and erythrocytes, while in the third case there was a slight decrease. In all 3 cases during administration of cortisone, there were slight increases in glutathione in whole blood and erythrocytes. Although abnormal glucose tolerance curves were observed in 2 cases during administration of cortisone, adequate data are not available for a significant correlation of changes in concentration of glutathione in whole blood and erythrocytes with alterations in glucose tolerance. Conn, Louis, and Johnston (1949) have observed a decrease in blood glutathione in association with impairment of glucose tolerance due to ACTH.

Carbohydrate Tolerance

That the administration of cortisone to man might cause impairment of tolerance for carbohydrate is suggested by several experimental and clinical observations: (1) cortisone and ACTH are capable of inducing hyperglycemia and glycosuria in rats (Ingle, 1941; Ingle, Li, and Evans, 1946); (2) impairment of carbohydrate tolerance is a frequent manifestation of Cushing's syndrome and is apparently directly related to overproduction of carbohydrate-active adrenal steroids in this condition (Sprague, Power, and Mason, 1949); (3) cortisone lessens sensitivity to insulin in patients having Addison's disease, greatly intensifies the diabetes of patients who have coexisting Addison's disease and diabetes mellitus (Sprague, Power, Mason and Cluxton, 1949), and, if the observations in the case reported by Boland and Headley (1949) prove to be generally true, augments the requirement for insulin of diabetic patients receiving cortisone.

Inasmuch as the doses of cortisone administered to patients who were subjects of this study were relatively small compared to those with which Ingle (1941) induced frank diabetes in animals, it would be anticipated (assuming an analogy between the carbohydrate metabolism of man and rat) that any impairment of carbohydrate tolerance occurring in the subjects of our study would be slight.

This reasoning proved to be correct. Although slight inconstant increases in the fasting blood sugar were observed in some cases during administration of cortisone or ACTH, the values never exceeded the normal range. In no case were more than traces of sugar in the urine observed.

Since glucose tolerance tests are somewhat more sensitive than determinations of the fasting blood sugar for detection of impairment of carbohydrate tolerance, they were performed in a small group of cases. None of the patients was known to have diabetes. Most of the tests were performed during administration of cortisone, but a few were performed before and after its administration as well. The data indicate that cortisone induced measurable impairment of carbohydrate tolerance in some cases: abnormal glucose tolerance curves, presumably due to cortisone, were observed in 4 cases (that is, cases 1, 11, 12, and 14) out of 12 in which the tests were performed (Table 2). In 3 (cases 1, 12, and 14) of these 4 cases, data on glucose tolerance

TABLE 2

Results of Glucose Tolerance Tests Performed before, during, and after Administration of Cortisone Acetate

Case	Family History of Diabetes	Before Cortisone	During Cortisone			After Cortisone	
		Blood sugar, mg per 100 cc	Days given (consecutive)	Average dose, mg	Blood sugar, mg per 100 cc	Days since last dose	Blood sugar, mg per 100 cc
1	No		169	90	88, 135, 185, 85		
9	No	90, 195, 230*	15	110	88, 173, 149*		
10	No	83, 149, 153*	13	154	83, 120, 153*		
11	Yes	78, 125, 103, 86	7	129	86, 149, 144, 86		
			16	200	98, 142, 153, 81	18	74, 102, 88, 60
12	No		7	129	88, 163, 86, 74		
			56	105	83, 163, 149, 53	128	90, 135, 63, 64
13	No	83, 135, 63, 66	9	122	85, 163, 85, 56		
14	Yes		15†	200	86, 103, 147, 103		
15	No		60	131	86, 113, 107, 72		
23	No	88, 164, 132, 87	11	118	78, 125, 115, 74		
			36	106	78, 115, 93, 81		
			12	100	86, 120, 111, 100		
25	No		11	127	78, 142, 100, 60		
26	No					20	78, 115, 86, 60
32	No		96	112	80, 141, 54, 78		
			162	107	78, 128, 74, 70		

* Exton and Rose's glucose tolerance test. Other tests were of "standard" type, employing 1 g of glucose per kilogram of body weight, and determining the blood sugar at 0, ½ or 1, 2, and 3 hours.

† This patient received 10 mg of estrone daily in addition to cortisone.

had not been obtained prior to administration of cortisone. In those cases in which the test was performed after stopping administration of the hormone (cases 11 and 12), it was found that glucose tolerance had reverted to normal.

Reference to Table 2 will show that in some instances (particularly cases 15, 23, and 32) no impairment of carbohydrate tolerance was demonstrable after protracted administration of cortisone. It will be noted for example that in case 32 no impairment could be demonstrated after administration of the hormone for as long as 162 days. In case 11, on the other hand, impairment was present after only seven days' administration.

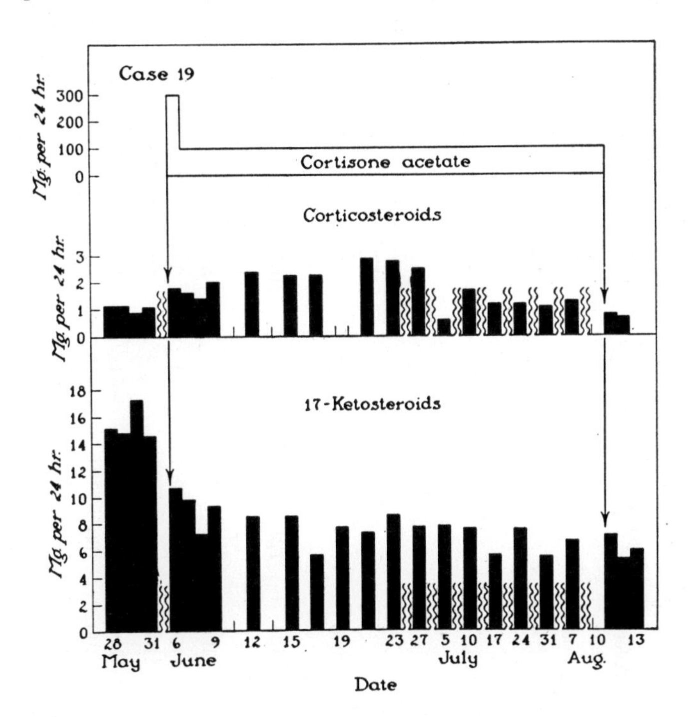

FIG. 6 (case 19). Urinary corticosteroids and 17-ketosteroids before, during, and after administration of cortisone for sixty-six days. Note prompt decrease in 17-ketosteroids when administration of hormone was begun.

EVIDENCES OF DEPRESSION OF ADRENAL CORTICAL FUNCTION BY CORTISONE

When cortisone or other adrenal hormones which have similar physiologic activity are administered to animals, atrophy of the adrenal cortices eventually ensues. Likewise, atrophy of the cortex of one adrenal gland occurs in patients who have a hyperfunctioning tumor of the cortex of the contralateral gland, particularly when the hormonal activity of the tumor is such as to induce Cushing's syndrome.

Such observations suggest that cortisone might cause atrophy of the adrenal cortices in the human being, and that depression of adrenal cortical

TABLE 3

Changes in Eosinophil Count in Response to 25 mg of ACTH before and during Administration and after Withdrawal of Cortisone

Case	Before Cortisone: Eosinophils per cu mm, 8 A.M.	12 M	% change	During Cortisone: Time given, days (consecutive)	Average daily dose, mg	Eosinophils per cu mm, 8 A.M.	12 M	% change	Time given, days (consecutive)	Average daily dose, mg	After Cortisone: Days since last dose	Eosinophils per cu mm, 8 A.M.	12 M	% change
											33	556	403	−28
3									73	100	86	425	145	−66
6	148	35	−76											
											8	66	65	−2
12									57	100	129	253	35	−86
14	62	6	−90	18	200	9	3	−66	18	200	18	26	9	−65
	38	9	−76											
16				145	124	121	84	−31	161	117	10	153	18	−88
19									66	103	9	376	334	−11
21				35	100	210	160	−24	56	100	11	121	115	−5
22	616	123	−80	20	100	229	257	+12	32	100	10	151	73	−52
				32	100	357	341	−4	32	100	56	473	163	−66
23	163	31	−80	20	100	223	103	−54	30	100	19	73	0	−100
				29	100	160	21	−87*						
				15	100	125	156	+25						
32									166	107	6	138	68	−5
									166	107	27	163	76	−53

* Urticarial reaction after ACTH.

function might be apparent for an interval following cessation of administration of the hormone, particularly if the hormone had been administered in large doses or for a prolonged period. We have, in fact, observed three types of evidence suggestive of adrenal depression by cortisone, as follows.

Asthenia Following Withdrawal of Cortisone. Beginning a few days after administration of cortisone was stopped, several patients complained of weakness of varying degree. The asthenia persisted a few days to several weeks. It was not consistently related to recurrence of articular disease following withdrawal of cortisone. Although it could not be attributed with certainty to adrenal cortical insufficiency, no other basis for it was apparent.

Urinary Steroids during Administration and after Withdrawal of Cortisone. It was noted repeatedly that there was a prompt decrease in urinary excretion of 17-ketosteroids soon after administration of cortisone was started, and persistently low levels of 17-ketosteroids and corticosteroids following cessation of prolonged administration of the hormone. These observations would seem to indicate depression of at least some functions of the adrenal cortex. Data in 1 case are shown in Fig. 6.

Response of the Adrenal Cortices to ACTH. A series of tests of adrenal response to a single 25-mg dose of ACTH, as measured by change in the number of circulating eosinophils four hours after injection of the hormone, was performed, according to the method of Thorn and his associates (1948). Results are summarized in Table 3. A normal response is defined [on the basis of the experience of Thorn and his associates (1948) and that of Balfour (1949)] as a 62% or greater decrease in the number of circulating eosinophils. In all of the 5 tests prior to administration of cortisone the response to ACTH was normal. Response to ACTH during administration of cortisone was normal in two tests and abnormal in six. The two "normal" responses are open to question because of the extremely low initial eosinophil count in 1 case (case 14) and the occurrence of a moderately severe urticarial reaction to ACTH in the other (case 23). After administration of cortisone had been stopped for varying periods of time, the response to ACTH was variable, being normal six times and abnormal seven times. It is perhaps significant that in cases 3, 12, and 22 the response changed from abnormal to normal as the interval after the last dose of cortisone increased. Thus, if one can assume that cortisone does not induce a change in the eosinophils themselves which makes them less responsive to adrenal cortical steroids, the overall results of these tests indicate that cortisone induced a diminished responsiveness of the adrenal cortices to ACTH.

The three lines of evidence suggesting depression of adrenal cortical function which have been cited, together with evidence from animal experiments, create a strong presumption that such functional depression may actually occur. Larger doses of cortisone might be expected to produce more significant depression. It is not likely to present a serious practical problem, however, except under conditions of stress, such as fever or trauma, when adrenal function might conceivably be inadequate.

Comment

The data which have been presented, while not constituting an exhaustive investigation of the physiologic potentialities of ACTH and cortisone, serve to indicate that they are powerful hormonal agents which are capable of influencing many metabolic processes. Their ability, in the human subject, to increase the catabolism of protein, influence the metabolism of electrolytes, and impair tolerance for carbohydrate has been demonstrated by these studies and those of others. As would be anticipated, there is evidence that cortisone may depress at least some functions of the adrenal cortices. Still other physiologic effects in human subjects are described elsewhere.

Our observations are presented with the thought that they may provide some guidance for further physiologic studies in human subjects, and also because the effects observed must receive some consideration in the eventual application of these hormones in the treatment of disease. Unless we are guilty of failing to appreciate some important meaning in our data, it can be said that they shed no light on the mechanism of action of these agents in rheumatic disease.

It is perhaps unnecessary to emphasize here that the varied physiologic alterations which may be induced by these hormones are more properly designated as "effects" than as "side effects" or "toxic reactions." In other words, the effects observed are exaggerated manifestations of the normal biologic activity of the hormones.

Summary

Metabolic studies of a group of patients, most of whom had rheumatoid arthritis, who received cortisone or ACTH or both for varying periods have been reported. One of the patients also received 17-hydroxycorticosterone (compound F). The studies were concerned chiefly with the effects of fairly prolonged administration of the hormones.

ACTH in a dose of 100 or 105 mg daily for twelve days induced a negative balance for nitrogen and potassium. There was initially a marked retention of sodium and chloride, and then, while the hormone was still being administered, increased excretion of these ions. A hypochloremic, hypokaliemic alkalosis developed, accompanied by some lowering of the plasma sodium.

Cortisone in a dose of 100 mg daily for twelve to thirty days induced only minimal alterations, or no alteration at all, in the balances for nitrogen, calcium, phosphorus, sodium, potassium, and chloride, and in the concentrations of the plasma electrolytes.

Cortisone in a dose of 200 mg daily for twelve to eighteen days regularly induced a negative balance for nitrogen and potassium. The effects on excretion of sodium and chloride were variable; usually there was retention of these ions early in the period of administration of the hormone, followed later by increased excretion. At this dose level, a hypochloremic, hypokaliemic alkalosis developed, similar in all respects to that observed in some cases of Cushing's syndrome. The concentrations of the plasma electrolytes gradually reverted to normal after withdrawal of the hormone.

The metabolic effects of cortisone in a woman in whom artificial menopause had been induced four years previously were not modified by the prior and simultaneous administration of 10 mg of estrone daily. In this case the simultaneous administration of 25 mg of testosterone propionate daily with the cortisone resulted in a decrease in the excretion of nitrogen and minimized loss of potassium, but did not prevent the development of a hypochloremic, hypokaliemic alkalosis.

The metabolic changes induced by the administration of compound F (900 mg in twelve days), as far as they were measured in 1 case, were not pronounced. Alkalosis of mild degree, associated with a slight decrease in plasma chlorides, developed.

Cortisone and ACTH increased excretion of creatine and uric acid in the urine. The increase in urinary creatine was prevented in part in 1 case by simultaneous administration of testosterone propionate.

Administration of cortisone induced slight increases in the concentration of serum albumin and more marked decreases in the concentration of serum globulin. These changes were most marked when the pretreatment values for serum albumin were depressed and those for serum globulin were elevated. Zinc sulfate turbidity of serum decreased significantly during administration of cortisone.

There was no consistent pattern of change in glutathione in whole blood and erythrocytes during administration of cortisone and ACTH in the 3 cases in which glutathione was determined. Adequate data are not available for a significant correlation of changes in concentration of glutathione in whole blood and erythrocytes with alterations in glucose tolerance.

The ability of cortisone to depress glucose tolerance was demonstrated in some cases. In others no impairment of glucose tolerance could be demonstrated even after prolonged administration of the hormone.

Evidence was obtained that cortisone is capable of depressing the function of the adrenal cortices in man as it does in the rat.

No correlation was recognized between the observed metabolic effects of cortisone, ACTH, and compound F and the favorable modification of the course of rheumatoid arthritis produced by these hormones.

REFERENCES

Balfour, W. M., unpublished data.

Boland, E. W., and Headley, N. E., Effects of cortisone acetate on rheumatoid arthritis. *J. Am. Med. Assoc.,* **141,** 301-308 (1949).

Conn, J. W., Louis, L. H., and Johnston, Margaret W., Metabolism of uric acid, glutathione and nitrogen, and excretion of "11-oxysteroids" and 17-ketosteroids during induction of diabetes in man with pituitary adrenocorticotropic hormone. *J. Lab. Clin. Med.,* **34,** 255-269 (1949).

Hench, P. S., Kendall, E. C., Slocumb, C. H., and Polley, H. F., The effect of a hormone of the adrenal cortex (17-hydroxy-11-dehydrocorticosterone: compound E) and of pituitary adrenocorticotropic hormone on rheumatoid arthritis; preliminary report. *Proc. Staff Meetings Mayo Clinic,* **24,** 181-197 (1949).

Ingle, D. J., The production of glycosuria in the normal rat by means of 17-hydroxy-11-dehydrocorticosterone. *Endocrinology,* **29,** 649-652 (1941).

INGLE, D. J., LI, C. H., and EVANS, H. M., The effect of adrenocorticotropic hormone on the urinary excretion of sodium, chloride, potassium, nitrogen and glucose in normal rats. *Endocrinology,* **39,** 32-42 (1946).

KUNKEL, H. G., Estimation of alterations of serum gamma globulin by a turbidimetric technique. *Proc. Soc. Exptl. Biol. Med.,* **66,** 217-224 (1947).

SPRAGUE, R. G., MASON, H. L., and POWER, M. H., Studies of the effects of adrenal cortical hormones on carbohydrate metabolism in human subjects. *Proc. Am. Diabetes Assoc.,* **9,** 149-169 (1949).

SPRAGUE, R. G., POWER, M. H., MASON, H. L., and CLUXTON, H. E., Metabolic effects of synthetic compound E (17-hydroxy-11-dehydrocorticosterone) in 2 patients with Addison's disease and in 1 with coexisting Addison's disease and diabetes mellitus. (Abstr.) *J. Clin. Invest.,* **28,** 812 (1949).

SPRAGUE, R. G., POWER, M. H., MASON, H. L., ALBERT, A., MATHIESON, D. R., HENCH, P. S., KENDALL, E. C., SLOCUMB, C. H., and POLLEY, H. F., Observations on the physiologic effects of cortisone and ACTH in man. *Arch. Internal Med.,* **85,** 199-258 (1950).

THORN, G. W., FORSHAM, P. H., PRUNTY, F. T. G., and HILLS, A. G., A test for adrenal cortical insufficiency; the response to pituitary adrenocorticotropic hormone. *J. Am. Med. Assoc.,* **137,** 1005-1009 (1948).

THE RESPONSE OF NEOPLASTIC LYMPHOID TISSUE TO INCREASED ADRENAL CORTICAL FUNCTION*

O. H. PEARSON, L. P. ELIEL, AND F. C. WHITE
Sloan-Kettering Institute, Memorial Hospital, New York

We have previously reported (Pearson *et al.*, 1949) that administration of adrenocorticotropic hormone (ACTH†) or cortisone acetate to patients with lymphoid tumors results in a marked and progressive decrease in the size of enlarged lymph nodes and spleens. These observations were made on four patients with chronic lymphatic leukemia, one with follicular lymphosarcoma, and one with Hodgkin's disease. In none of these patients was a complete clinical remission of the disease obtained, and regrowth of tumor masses occurred in all patients within ten days to three months after discontinuing the hormones.

In this paper the metabolic changes produced by the administration of ACTH to one patient with chronic lymphatic leukemia are presented. Data obtained on the phosphorus content of tumor tissue and the excretion of phosphorus during the administration of ACTH provide evidence that tumor tissue was destroyed. Increasing the dietary intake during the period of ACTH administration produced a marked shift in the nitrogen and phosphorus balance indicating that the metabolic response induced by ACTH may be markedly influenced by the dietary intake.

The patient is a 57-year-old white male with chronic lymphatic leukemia of one and one-half years known duration. His presenting complaint was enlargement of cervical nodes. Examination revealed generalized lymphadenopathy and splenomegaly. Blood studies showed a white blood count of 200,000 of which 99% were mature lymphocytes. Treatment with x ray and radioactive phosphorus during a nine-month period resulted in slight, transient improvement; but further radiation therapy was then discontinued because of purpuric manifestations and low platelet counts. During the nine-month period prior to the onset of this study no therapy was given except blood transfusions.

This experiment was carried out on a metabolic ward in an air-conditioned room, and the patient was fed a constant, accurately measured diet. Balance determinations were made of nitrogen, phosphorus, calcium, potassium, sodium, and chloride. The urinary excretion of uric acid, creatine, and creatinine was measured. Urinary steroid studies were carried out by Dr. Konrad Dobriner and are reported by him in detail in his paper.

---

* This study was supported by grants from the U. S. Public Health Service, the Office of Naval Research, the Atomic Energy Commission, the Damon Runyon Cancer Research Fund, and the American Cancer Society.

† The ACTH used in this study was generously supplied by Dr. John R. Mote of the Armour Company.

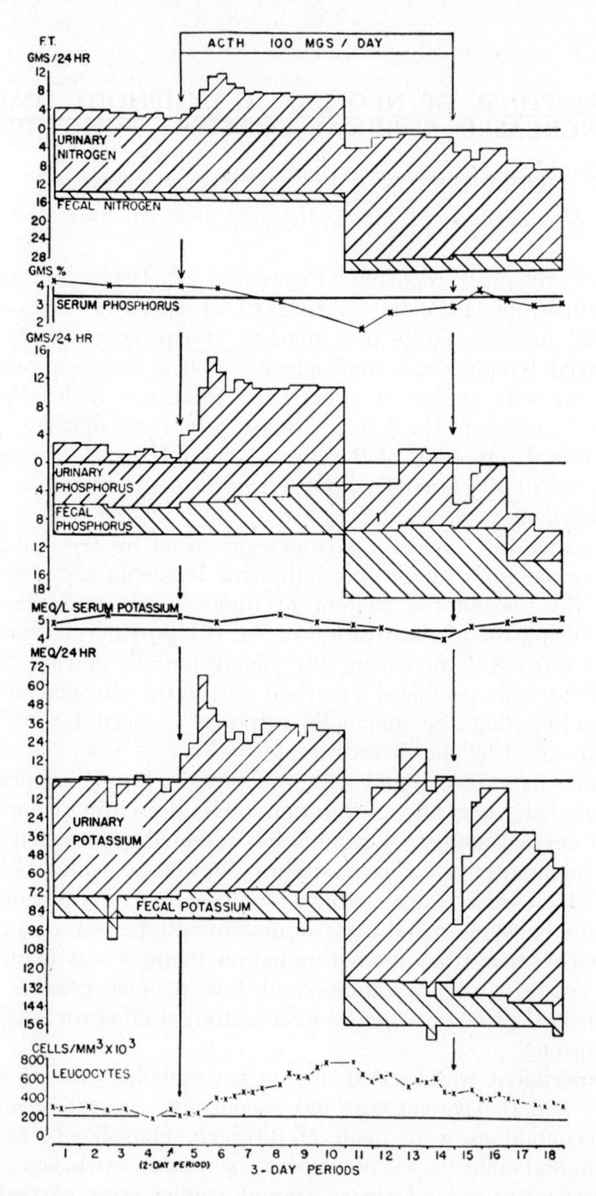

FIG. 1. Patient F. T. M. H. No. 95513. Male. Age 57. Chronic lymphatic leukemia. Effects of ACTH on nitrogen, phorphorus, and potassium balances and on white blood count.

Note. The scale for nitrogen has been selected to be 15 times the scale for phosphorus which is the approximate ratio of these elements in normal protoplasm.

After 14 days of control observation, ACTH was administered for 30 days in a dosage of 100 mg daily in four divided doses at intervals of 6 hours. While ACTH was given there was a marked and progressive decrease in the size of enlarged lymph nodes and of the spleen. During the 12-day control period after stopping ACTH there was rapid regrowth of lymph nodes and enlargement of the spleen. Because of hunger, the dietary intake was approximately doubled on the nineteenth day of ACTH, and this diet was maintained for the remainder of the study.

The data are presented in Fig. 1. The balance determinations of nitrogen, phosphorus, and potassium are plotted according to the recommendations of Reifenstein *et al.* (1945). In the initial control period, there was a negative nitrogen and phosphorus balance, and the amount of phosphorus lost is approximately equal to the amount of nitrogen lost. During the first 18 days of ACTH administration there is a markedly negative nitrogen and phosphorus balance, and there is approximately twice as much phosphorus lost as would be anticipated from the nitrogen loss. In the last 6 days of the after control period, there is a markedly positive nitrogen and phosphorus balance, and the retention of phosphorus is proportionately greater than the retention of nitrogen.

Biopsies of muscle and lymphoid tumor masses were obtained before and during the administration of ACTH and were analyzed for nitrogen and phosphorus content. There was no significant change in the tissue content of nitrogen and phosphorus before and after therapy. The phosphorus content of tumor tissue per unit of nitrogen is approximately three times that of muscle. This has also been observed in 3 other patients (Eliel, Pearson, Katz, and Kraintz, 1950).

These data indicate that the large loss of phosphorus during ACTH administration and the large retention of phosphorus during the latter half of the after-control period are best explained by loss and gain, respectively, of tumor tissue which has a high phosphorus content.

On the nineteenth day of the ACTH period the dietary intake was approximately doubled. This was associated with an abrupt change from negative nitrogen and phosphorus balance to slightly positive balance. It was noted clinically during this period that little change occurred in the size of lymph nodes and spleen. There was no remarkable change in the urinary steroid excretion associated with the increase in dietary intake. These data indicate that the effect of increased adrenal cortical function on tissue anabolism or catabolism may be modified by the level of the dietary intake. Since all the constituents of the diet were increased in this study, there is no clue as to what dietary substances are most important in exerting this influence. Ingle (1949) has presented evidence that the level of protein intake may be of primary importance.

The potassium balance follows quite closely that of phosphorus. The loss of potassium during the first 18 days of the ACTH period and the retention of potassium during the last 6 days of the final control period are greater than those anticipated from the nitrogen balance using the accepted ratio of nitrogen to potassium in protoplasm (Reifenstein *et al.,* 1945).

Analysis of tumor and muscle tissue in this patient for potassium reveals a higher content of potassium per unit of nitrogen in tumor as compared to muscle (tumor, K/N = 4.05 meq/1 g; muscle, K/N = 2.64 meq/1 g). It would appear that the larger than anticipated loss and gain of potassium may be related to the loss and gain of tumor tissue with a higher content of potassium. The large retention of potassium and phosphorus during the first 3 days of the after-control period cannot be accounted for on the basis of tissue anabolism.

The excretion of uric acid, creatine, and urinary steroids increased during the administration of ACTH . There was retention of sodium and chloride during the first 27 days of the ACTH period which was more pronounced when the sodium chloride intake was increased. The retention of sodium and chloride was associated with the development of severe edema and gain in body weight. Some diuresis of sodium and chloride began 3 days before ACTH was stopped, but was not extensive until after ACTH was discontinued. The urinary excretion of calcium increased progressively during administration of ACTH and decreased when ACTH was stopped. The calcium balance remained negative throughout the study varying from —100 to —300 mg per day. The white blood cell count rose from an initial level of 250,000 to a peak of 750,000 after 18 days of ACTH and then gradually receded to the initial level.

Summary

A metabolic study of the effects of increased adrenal cortical function induced by the administration of ACTH to a patient with chronic lymphatic leukemia is presented. Data obtained on the phosphorus content of tumor tissue and the excretion of phosphorus during the administration of ACTH provide evidence that tumor tissue was destroyed. Increasing the dietary intake during the period of ACTH administration produced a marked shift from a negative to a positive balance of nitrogen and phosphorus, indicating that the effect of increased adrenal cortical function on tissue anabolism or catabolism may be modified by the level of the dietary intake.

References

Pearson, O. H., Eliel, L. P., Rawson, R. W., Dobriner, K., and Rhoads, C. P., ACTH- and cortisone-induced regression of lymphoid tumors in man. *Cancer,* **2,** 943 (1949).

Reifenstein, E. C., Albright, F., and Wells, S. L., The accumulation, interpretation, and presentation of data pertaining to metabolic balances, notably those of calcium, phosphorus, and nitrogen. *J. Clin. Endocrinol.,* **5,** 367-395 (1945).

Eliel, L. P., Pearson, O. H., Katz, B., and Kraintz, F. W., A comparison of lymphoid tumor and muscle electrolyte composition in patients treated with ACTH and cortisone acetate. *Federation Proc.,* **9,** 168 (1950).

Ingle, D. J., Some studies on the role of the adrenal cortex in organic metabolism. *Ann. N. Y. Acad. Sci.,* **50,** 576-595 (1949).

ADRENAL FUNCTION IN THE NEWBORN*

ELEANOR H. VENNING
McGill University Clinic, Royal Victoria Hospital, Montreal, Canada

The rapid changes occurring in the adrenal gland of the newborn infant have been the subject of numerous investigations. In an attempt to learn something about the function of this gland at this time urinary corticoids and 17-ketosteroids have been measured in full-term and premature infants. The effect of stress and the administration of adrenocorticotropic hormone upon the excretion of these adrenal metabolites has also been studied.

At birth, the cortex is composed of two zones, an outer or true cortical part and an inner zone, the fetal cortex. According to Benner (1940) the fetal cortex is present as a definite structure by the fourth fetal month. It grows rapidly throughout the last trimester of fetal life and is characterized by increasing hyperemia and infiltration of lipoid material. At birth this zone begins to degenerate and involution proceeds rapidly for a period of two to three weeks. The adult cortex grows steadily throughout infancy. This growth first replaces the destroyed fetal cortex and later is manifested by an increase in the size of the adrenal gland. Within a month after birth the adult or true cortex has doubled its size.

Several theories have been presented to explain these morphological changes such as alterations in oxygen tension (Goldzieher, 1934) and tissue respiration which occurs at the time of birth. Grollman (1936), Broster (1937), and others have suggested that this structure has an androgenic function, but attempts by Gersh and Grollman (1939) and by Carnes (1940) to demonstrate the presence of androgenic substance in the fetal adrenal have failed.

In this paper two separate studies on adrenal cortical function of the newborn will be reviewed. The first one is concerned with the excretion of biologically active corticoids, the glucocorticoids, in a group of full-term and premature infants at varying ages and has been reported by Venning, Randall, and György (1949). The second deals with the excretion of corticoids (measured by a chemical procedure) and 17-ketosteroids in a series of full-term infants and the effect of adrenocorticotropic hormone upon these metabolites. This later investigation was carried out by Read, Venning, and Ripstein (1950).

Methods

Glucocorticoids. The glucocorticoids were measured by the bioassay of Venning, Kazmin, and Bell (1946) which is based upon liver glycogen deposition in adrenalectomized mice. The standard used was cortisone.

* Supported in part by a grant from the National Research Council, Canada.

Chemical Corticoids. The chemical method used for the assay of corticoids was that of Daughaday, Jaffe, and Williams (1948). The neutral water soluble fraction was used for this determination and the standard was desoxycorticosterone.

17-Ketosteroids. The 17-ketosteroids were determined by the procedure of Callow *et al.* (1938), and the correction factor of Gibson and Evans (1937) was applied for the elimination of interfering chromogens.

Eosinophil Counts. Eosinpohil counts were performed on capillary blood using the method of Randolph (1944).

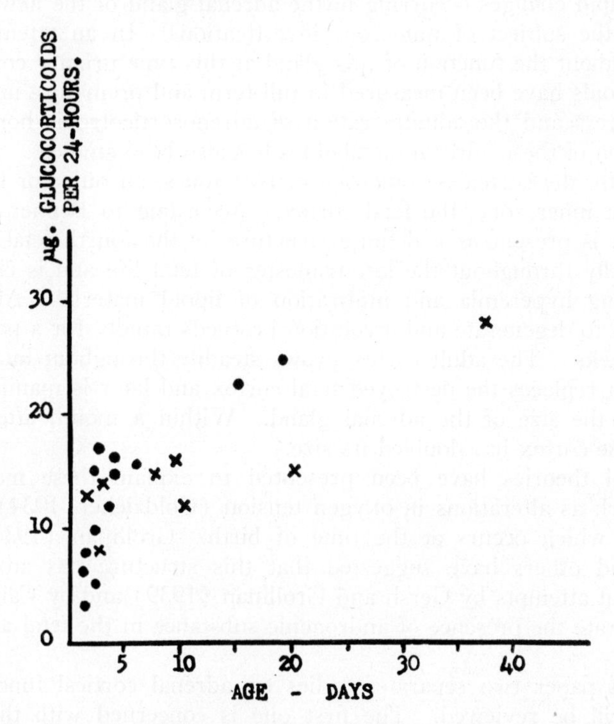

Fig. 1. Excretion of glucocorticoids in full-term and premature babies.
●—Full term X—Premature

Excretion of Glucocorticoids in Newborn Infants

In order to obtain sufficient urinary material for biological assay, collections of urine were made over periods of 3 to 4 days. Only male babies were used in these studies. In ten healthy full-term infants, the amount of glucocorticoids found in the urine during the first 5 days of life ranged from 4 to 17 µg per 24 hours (Fig. 1). The average value was 11 µg per 24 hours. In three premature babies of the same age, the values ranged from 8 to 14 µg per 24 hours. These infants were in good health at the time of the assays and their prematurity varied from four to six weeks. The values obtained on these infants were within the range observed for the full-term babies. Several assays were carried out on full-term and premature babies between the tenth and thirty-seventh day of life, and results indicated that after the fifteenth day there was a gradual increase in the excretion of corticoids with advancing age of the baby. No difference could be detected between the full-term infant and the premature infant. Thus there is a slowly rising excretion of corticoids at a time when rapid degeneration of the fetal cortex is occurring. These findings would be against any suggestion that the fetal cortex is concerned with the elaboration of corticoids. On the other hand, the steady increase in output of glucocorticoids parallels the growth of the true or adult cortex.

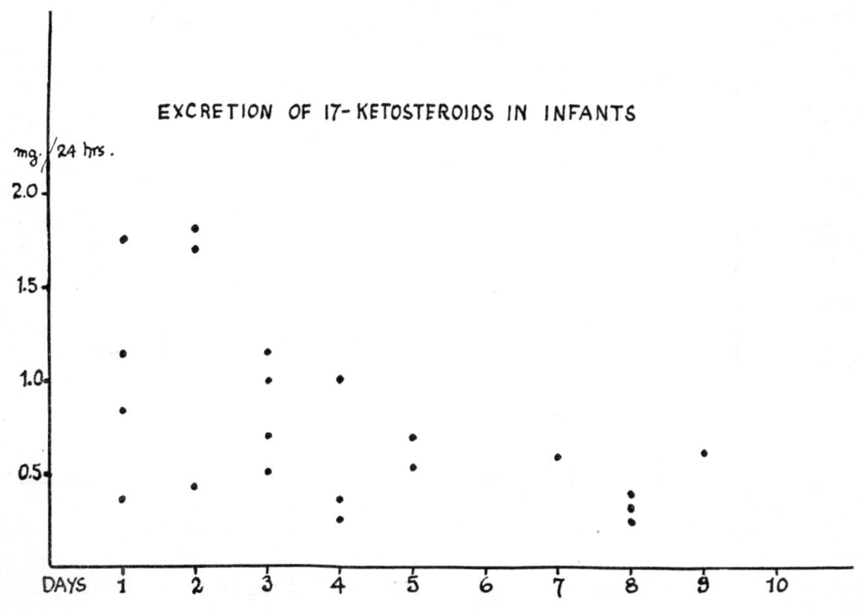

Fig. 2. Excretion of 17-ketosteroids in four infants.

Excretion of 17-Ketosteroids

The excretion of 17-ketosteroids has been followed in a few full-term male infants. Although only very small amounts of 17-ketosteroids are found in the urine of young children, assays carried out on the urine of newborn infants suggested that somewhat greater amounts are excreted during the first few days of life than in the second week of life.

Four infants were studied and in two of these infants the urinary 17-ketosteroids were in the order of 1.7 and 1.8 mg per 24 hours at 2 days of age. Over a period of one week they gradually decreased to values below 0.5 mg per 24 hours (Fig. 2). These findings are of interest because Grollman and Broster have suggested that the fetal cortex might be composed of androgenic tissue. The possibility that the increased excretion of 17-ketosteroids might be due to the passage of these substances through the placenta from the maternal circulation must be considered. A larger series of observations will have to be made before any conclusions may be drawn as to the origin of these substances.

Excretion of Corticoids (Chemical Method) in the Newborn

In another series of newborn infants the urinary corticoids were determined by means of the chemical method of Daughaday *et al.* (1948) which measures the amount of formaldehyde liberated following oxidation with periodic acid. These studies were carried out in collaboration with Dr. Charles Read. Only male infants were used. The urine was collected during the first 3 days of life and again between the eigth and eleventh days. The values are charted in Fig. 3. In eight infants, during the first 3 days of life,

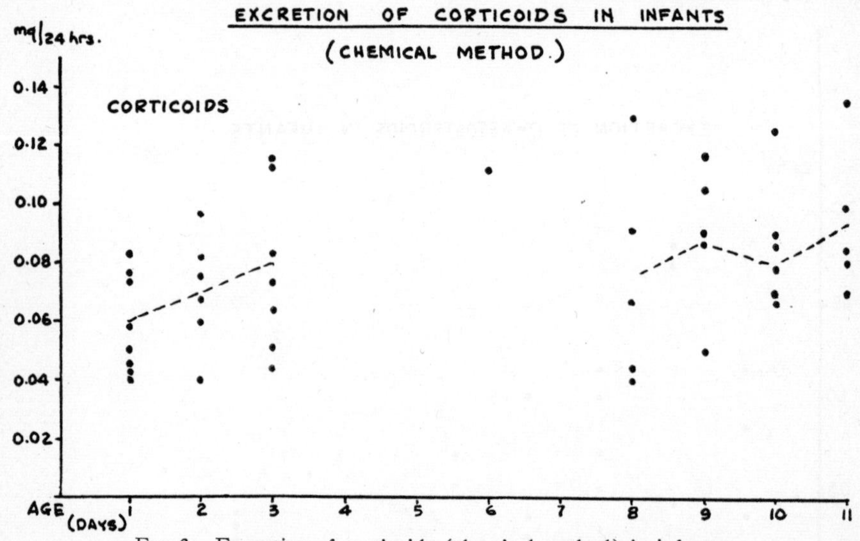

FIG. 3. Excretion of corticoids (chemical method) in infants.

the urinary corticoids ranged from 0.040 mg to 0.115 mg per day. The average output for these 3 days was 0.070 mg. In the second week of life the values ranged from 0.040 to 0.133, the average being 0.083 mg per 24 hours. A statistical analysis of these results indicated that the difference was just barely significant. In another group of infants, a rise in glucocorticoids was observed after the fifteenth day. Other investigators such as Day (1948) and Matson and Longswell (1949) have reported values on the excretion of corticoids in newborn infants by other methods. Their data showed no evidence of decreasing excretion of these substances with increasing age.

Effect of Adrenocorticotropin upon Urinary Corticoids

Two lots of adrenocorticotropin were used in this study, the Armour preparations 37-KE and H-2403.* It was administered intramuscularly and the dosage is expressed in terms of the Armour Standard La-I-A. After the injection the infants exhibited some pallor which lasted for approximately 3 hours. No other signs of posterior pituitary contamination were noted. Beginning with 0.3 mg the dose was gradually increased to 20 mg. Single

* The author is indebted to Dr. John R. Mote of the Armour Company for the adrenocorticotropin used in these studies.

injections were administered for amounts up to 5 mg. With higher dosages, multiple injections of 5 or 7 mg were given. No significant response in the output of urinary corticoids was observed until the dose reached a level of 5 mg. With this amount one infant showed a definite response, while another infant given a similar amount, failed to do so. With 14 and 20 mg administered in divided doses, significant increases in urinary corticoids were observed.

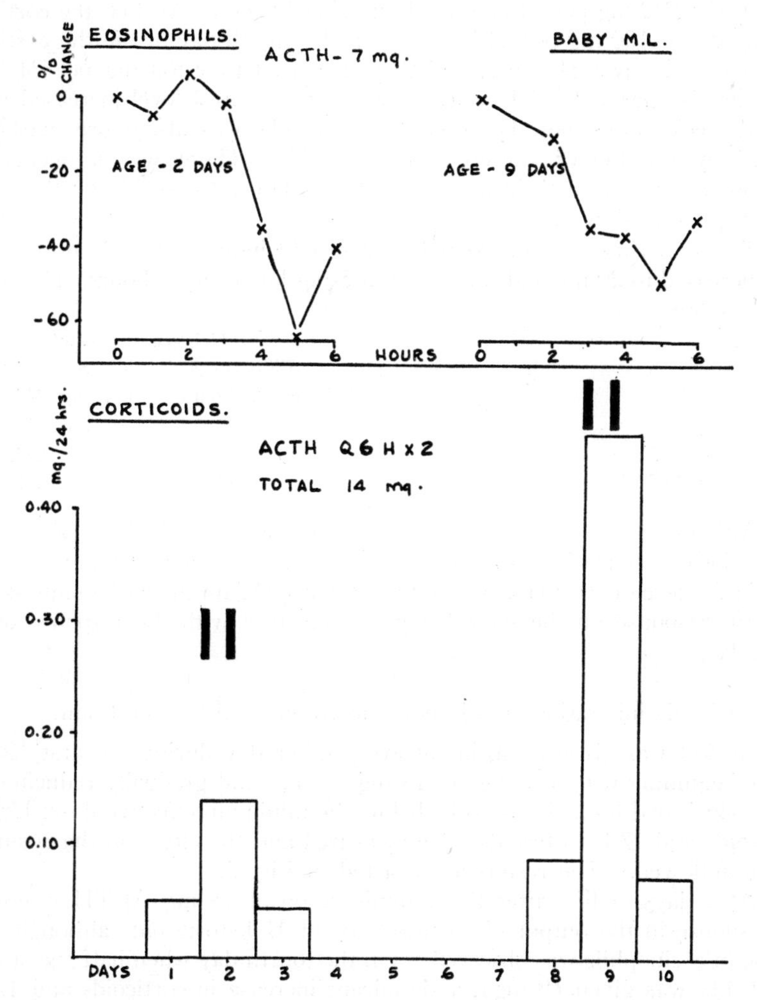

FIG. 4. Effect of ACTH on the excretion of corticoids and eosinophil counts in an infant on the second and ninth day of life.

Comparison of the Effect of Similar Doses of Adrenocorticotropin Given to Infants at 2 Days and at 9 Days of Age

As there is a higher mortality rate in infants during the first few days of life, it has been suggested that the adrenal might not be capable of responding adequately to stress at this time. In order to determine whether there was any difference in response at different ages, ACTH was given to two newborn infants on the second day of life, and this was again repeated one week later. In the first experiment 14 mg ACTH was administered (in divided doses) on the second and ninth days of life. The excretion of urinary corticoids and the hourly changes in eosinophil counts following the first injection of ACTH are shown in Fig. 4. On the first day of life, the excretion of urinary corticoids was 0.042 mg per 24 hours. Following 14 mg of ACTH, the corticoids increased to 0.130 mg per 24 hours. On the eighth day of life the corticoids were 0.090 mg per 24 hours, and they increased to 0.468 mg per 24 hours following 14 mg of ACTH. Not only was the control level increased in the second week of life but the response to ACTH was also more marked as judged by the higher excretion of corticoids. There was a decrease in eosinophil counts in both instances, but the response to ACTH on the second day of life was slower.

This experiment was repeated on a second infant. The dose of ACTH was increased to 20 mg and was given in 5-mg lots every 6 hours. The results are as follows:

	Urinary Corticoids	
	First Week mg/24 hours	Second Week mg/24 hours
Control	0.042	0.067
ACTH, 20 mg	0.259	0.332
After ACTH	0.143	0.115

Although there was an increase in excretion of corticoids over the first week, the difference in response was not as marked in this infant receiving the higher dose of ACTH as in the first infant. Again the eosinophils showed a lag in response on the second day as compared with the response on the ninth day.

Repeated Administration of Adrenocorticotropin to the Same Infant

ACTH was given to an infant every other day during the first 10 days of life beginning with a dosage of 18 mg ACTH and gradually reducing it to 3 mg which had been shown to be below the minimum effective dose. Urinary corticoids and 17-ketosteroids were measured and the effect on the eosinophil counts followed. The results are charted in Fig. 5.

At 2 days of life, after the administration of 18 mg ACTH, there was no response in the output of corticoids or of 17-ketosteroids although a decrease in eosinophils was observed. On the fourth day when half this amount of ACTH was given (9 mg), a significant increase in corticoids and 17-ketosteroids occurred. When the dose of ACTH was further reduced to 6 mg on the sixth day, the response was less. Two days later with 5 mg ACTH a significant increase in corticoids and 17-ketosteroids was again obtained. On the tenth day 3 mg ACTH failed to cause an increase of corticoids and 17-ketosteroids although there was a fall in eosinophils.

In this study an attempt was made to determine whether repeated administration of ACTH would alter the response of the adrenal to minimal doses of ACTH. The failure of this infant to respond to a relatively high dose of ACTH (18 mg) at two days of age with increased corticoids and 17-ketosteroid excretion and its ability to do so at the sixth and eighth day with much smaller amounts of ACTH would indicate that the adrenal gland at birth is somewhat refractory to stimulation with ACTH. The sensitivity of the adrenal appears to increase with advancing age of the infant. If this increase in sensitivity is due solely to repeated stimulation with ACTH then one would expect a response at 10 days of age with 3 mg of ACTH. The evidence obtained in these studies would indicate that the difference in sensitivity is a real effect.

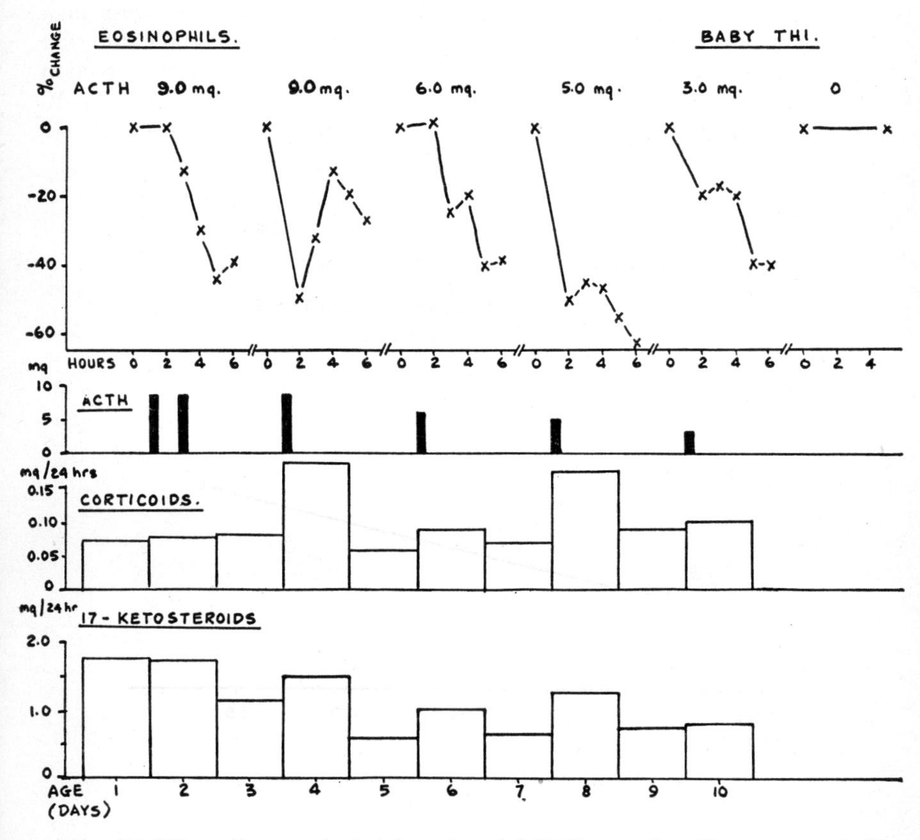

FIG. 5. Effect of repeated administration of ACTH on eosinophil counts, corticoids, and 17-ketosteroids in a newborn infant.

In this infant the slowly increasing output of corticoids can be observed over the period of study, while the 17-ketosteroids show an increased output at birth, decreasing to lower levels by the fifth day. In spite of this divergence in excretion rates of the two adrenal metabolites, when a rise in corticoids occurred after stimulation with ACTH, a similar increase was also observed in 17-ketosteroid excretion.

The decrease in eosinophil count appears to be a more sensitive index of adrenal stimulation than the excretion of steroids. It is difficult to evaluate this response in this infant as there was a decrease in eosinophil counts with all dosages of ACTH and no relationship between rate of fall of eosinophils and dosage of ACTH could be observed.

Effect of Trauma upon the Excretion of Glucocorticoids

It has been shown in adults that various types of damage such as operations, infections, and fractures will elicit a response in the adrenal as evidenced by an increase in excretion of glucocorticoids. That the pituitary and the adrenal of the newborn infant is also capable of responding to stress in a similar manner is indicated by the findings in two premature babies suffering from atelectasis. One of these, a male, was delivered by Caesarian section one month prior to term. The mother had been a severe diabetic for 10 years. At birth the baby was limp with a moderate amount of edema. He

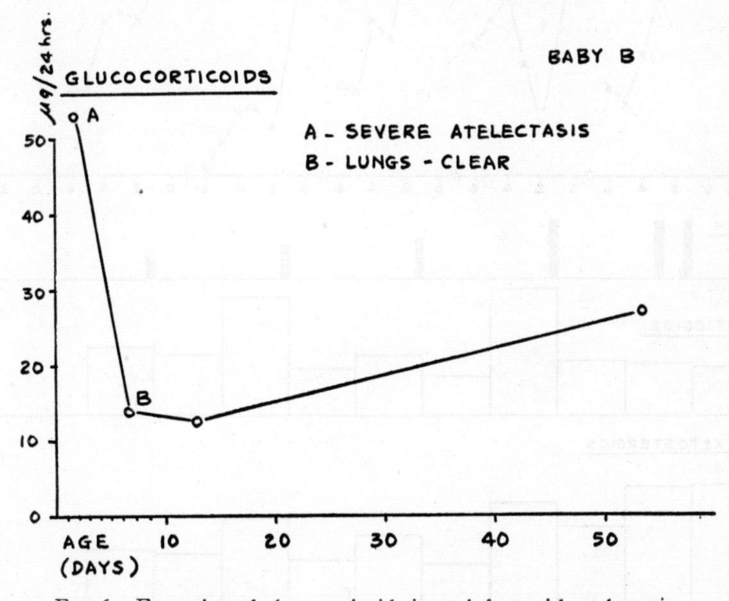

FIG. 6. Excretion of glucocorticoids in an infant with atelectasis.

was placed in an incubator and given oxygen. About one-half hour later, the baby became dyspneic and cyanosis was apparent if the oxygen was stopped. The clinical findings of complete collapse of the left lung with partial collapse of the right were confirmed by x ray. There was a slight improvement on the following day with some aeration of the right lung and gradual improvement from then on, but the lungs were not clear clinically and radiologically until the seventh day of life. Urine collection began at 6 hours after birth and was continued for 72 hours. It contained an equivalent of 53 μg of glucocorticoids per 24 hours, approximately five times the normal amount. Subsequent collections were made at the seventh, eleventh, and fifty-fourth days of life. As the condition of the baby improved, the urinary glucocorticoids returned to lower levels. The increased amount, 27 μg per 24 hours, found at the fifty-fourth day is within the normal range for that age. The results are charted in Fig. 6.

Conclusions

1. Newborn infants excrete small amounts of corticoids that can be detected by both bioassay and chemical methods.
2. After the second week gradual increase in the excretion of glucocorticoids is observed.
3. No difference could be detected between the full-term and the premature infant.
4. In some infants there is an increased excretion of 17-ketosteroids at birth. This gradually decreases to lower levels within 4 to 5 days.
5. The newborn infant responds to adrenocorticotropin in the same manner as the adult with fall in eosinophils and rise in urinary corticoids and 17-ketosteroids.
6. At birth the adrenal gland appears to be less responsive to stimulation with ACTH than in the second week of life.
7. The minimum effective dose of ACTH that would elicit a rise in urinary corticoids in the second week of life is in the order of 5 to 10 mg.
8. Under sufficient stress such as atelectasis, the newborn premature infant is capable of responding with an increased excretion of glucocorticoids.

REFERENCES

Benner, Miriam C., *Am. J. Path.*, **16**, 787 (1940).
Broster, L. R., *Arch. Surg.*, **34**, 761 (1937).
Callow, N. H., Callow, R. K., and Emmens, C. W., *Biochem. J.*, **32**, 1312 (1938).
Carnes, W. H., *Proc. Soc. Exptl. Biol. Med.*, **45**, 502 (1940).
Daughaday, W. H., Jaffe, H., and Williams, R. H., *J. Clin. Endocrinol.*, **8**, 166 (1948).
Day, E. M. A., *Med. J. Australia*, **2**, 122 (1948).
Gersh, I., and Grollman, A., *Am. J. Physiol.*, **126**, 368 (1939).
Gibson, J. G., Jr., and Evans, W. A. J., *J. Clin. Invest.*, **16**, 301 (1937).
Goldzieher, M. A., *Endocrinology*, **18**, 179 (1934).
Grollman, A., *The Adrenals*, Williams and Wilkins, Baltimore, 1936.
Matson, C. F., and Longswell, B. B., *J. Clin. Endocrinol.*, **9**, 646 (1949).
Randolph, T. G., *J. Allergy*, **1**, 89 (1944).
Read, C. H., Venning, Eleanor H., and Ripstein, M. P., *J. Clin. Endocrinol.*, **10**, 845 (1950).
Venning, Eleanor H., Kazmin, V. E., and Bell, J. C., *Endocrinology*, **38**, 79 (1946).
Venning, Eleanor H., Randall, J. Perlingiero, and György, Paul, *Endocrinology*, **45**, 430 (1949).

STEROID EXCRETION AND ADRENAL FUNCTION IN NEOPLASTIC DISEASE*

K. DOBRINER, S. LIEBERMAN, H. WILSON, B. EKMAN AND C. P. RHOADS

Sloan-Kettering Institute, Memorial Hospital, New York

Since 1940 the aim of our continuing investigation has been to understand the relationship between endocrine function and neoplastic disease (Dobriner, 1948; Dobriner and Lieberman, 1950). In patients with neoplastic growth the excretion of adrenal cortical and gonadal steroid metabolites is markedly diminished, and often certain steroids are no longer demonstrable (Fig. 1). If one assumes that the steroid excretion is a measure of hormone production by the adrenals and gonads, then one is forced to the conclusion that there is adrenal and gonadal dysfunction in malignancy. Confirmation of deranged adrenal function in neoplasia is afforded by the fact that 11-hydroxyetiocholanolone, a metabolite of adrenal cortical origin, is found in the urine of a significant number of patients with neoplastic disease whereas it is rarely present in the urine of normal individuals (Dobriner, 1948). Adrenal dysfunction existed in two patients several years before there was any obvious manifestation of the tumor as evidenced by the finding that 11-hydroxyetiocholanolone was consistently present before the tumor was diagnosed. In one of these patients the excretion of this abnormal metabolite has continued for more than five years after surgical removal of the breast without recurrence of the lesion. This indicates that adrenal function may be involved in the course or in the cause of neoplastic disease.

The question may then be asked whether stimulation of the adrenal gland by ACTH would restore the gland to a more nearly normal function and be reflected in an alteration in the course of the disease? There is evidence from the work of Heilman and Kendall (1944) that administration of cortisone decreases tumor growth in mice; furthermore, Dougherty and White (1943) have shown that adrenal hormones are extremely active agents in the dissolution of lymphatic tissue.

In order to study these questions in man we administered ACTH to a group of patients with cancer and lymphatic disease. This is a cooperative study with Drs. Pearson, Eliel, and Rawson, and the clinical and metabolic findings have been reported elsewhere (1949, 1950) as well as in this

* This investigation was aided by grants from the American Cancer Society (on recommendation of the Committee on Growth of the National Research Council), Ayerst McKenna and Harrison Ltd., the Jane Coffin Childs Memorial Fund for Medical Research, the Commonwealth Fund, the Anna Fuller Fund, the Lillia Babbitt Hyde Foundation, the Albert and Mary Lasker Foundation, and the National Cancer Institute, U. S. Public Health Service.

The ACTH used in this study was generously supplied by Dr. John R. Mote of the Armour Company.

volume. We shall report the effect of ACTH on hormone production as evidenced by the steroid metabolites in the urine. Figure 2 illustrates the steroid excretion before, during, and after administration of ACTH in a female patient with cancer of the breast (C. B.). The values shown were obtained from the crude neutral extracts before fractionation (Dobriner, Lieberman, and Rhoads, 1948; Lieberman and Dobriner, 1949). It is apparent that in this subject the administration of ACTH in dosage of 100 mg per day for 18 days resulted in only minor changes in the excretion of ketosteroid

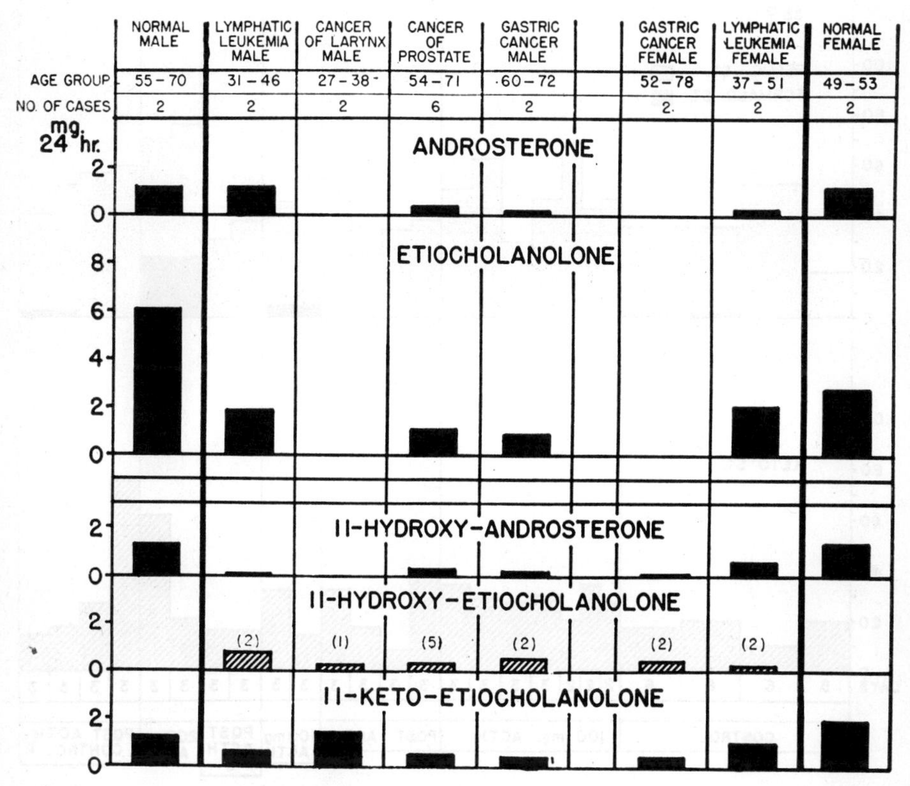

FIG. 1. A comparison of the excretion patterns of five α-ketosteroids of patients with neoplastic disease with those of normal men and women.

and reducing steroids. Following a 12-day control period without ACTH, 100 mg of ACTH per day was again given for 6 days, without any marked alteration in the excretion of the urinary steroids. When, however, 200 mg ACTH per day was given for a 6-day period, there was a pronounced increase in steroid excretion as measured by (1) the Zimmermann reaction, (2) by the total reduction, and (3) by the production of formaldehyde after periodate oxidation. Following cessation of ACTH the steroid excretion returned slowly to the control level. It is possible even with these very crude measurments to demonstrate that there was a marked change in steroid excretion when adequate amounts of ACTH were administered. Steroid excretion patterns were made during the periods where formaldehydogenic steroids were determined (cf. Fig. 2).

The effect of ACTH is more clearly apparent when individual steroids are examined. In Fig. 3 (C. B.) is shown the steroid excretion pattern of the α-ketosteroids in this subject; comparison is made with the pattern

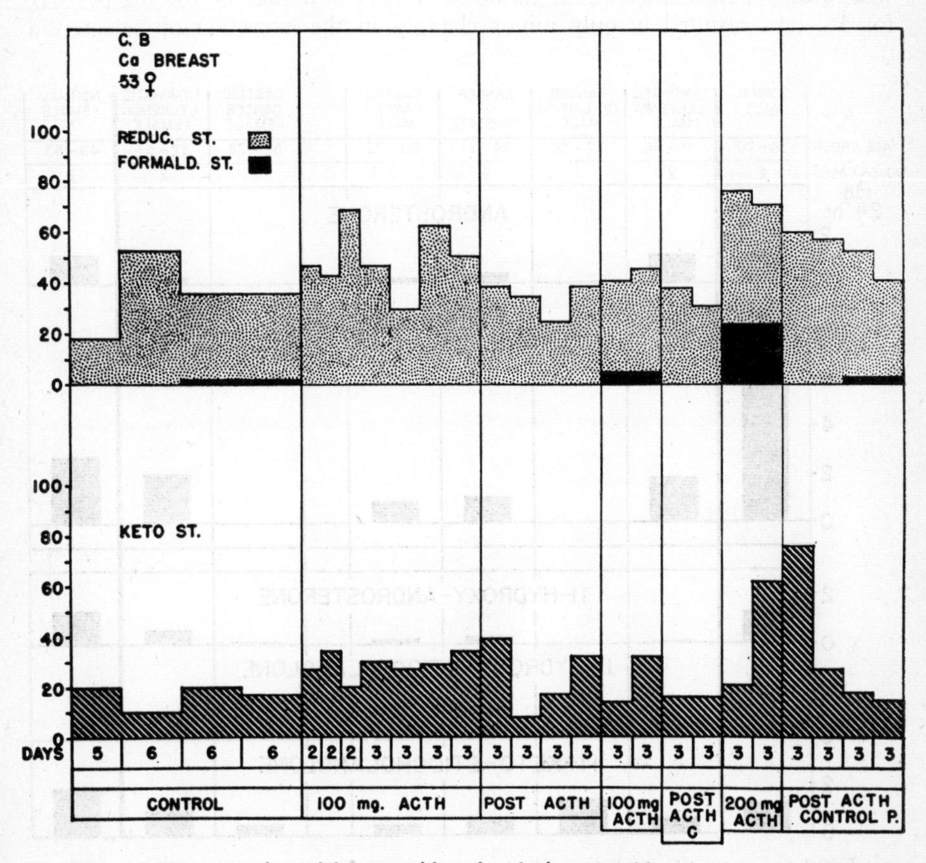

FIG. 2. The excretion of ketosteroids, of reducing steroids, and formaldehydogenic steroids by a woman with cancer of the breast before, during, and after the administration of ACTH.

of two normal females of similar age. The excretion of the five most abundant ketosteroids during the control period in the cancer patient was somewhat less than that of normal subjects. The effect of 100 mg of ACTH in this patient was slight, but there is an indication that these five compounds were increased to a level very similar to that of the normal controls. The pattern of the period during administration of 200 mg ACTH shows values somewhat above the normal range especially for etiocholanolone and pregnanolone. 11-Hydroxyetiocholanolone was present in the urine in the control period and

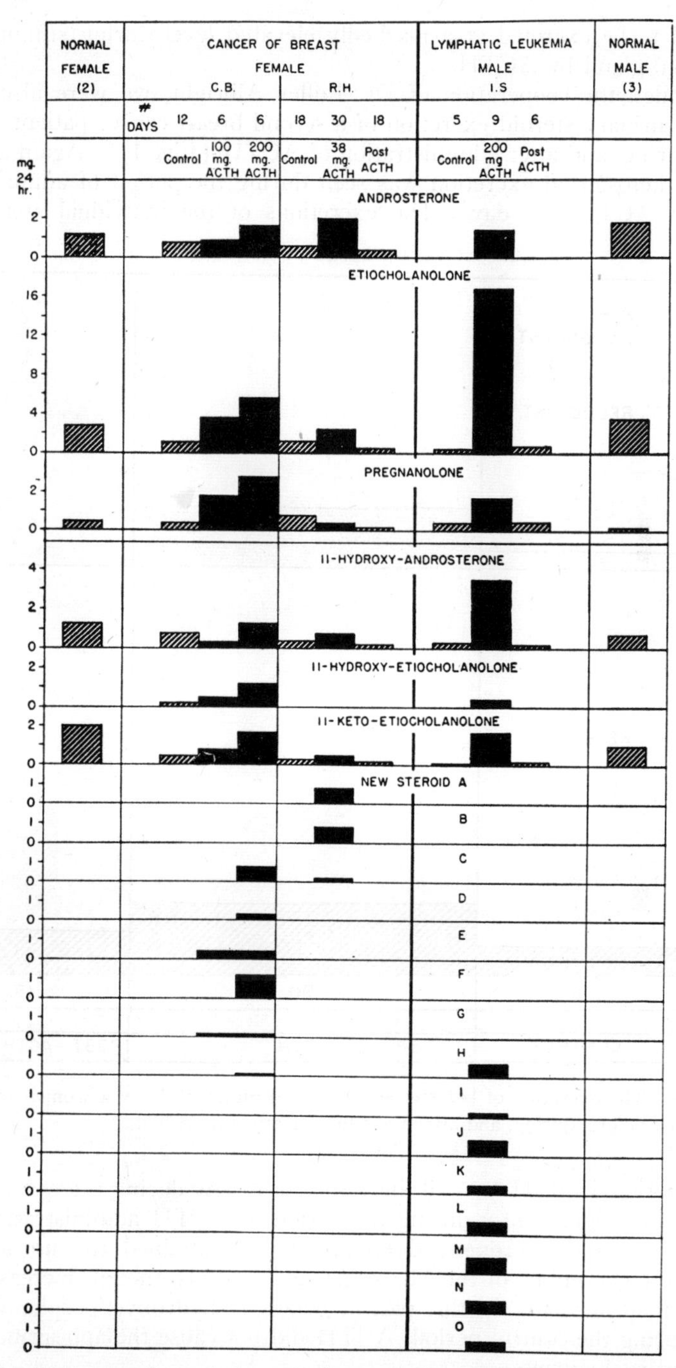

FIG. 3. The comparison of the excretion patterns of six known steroids and fifteen unidentified compounds of two women with cancer of the breast and a man with chronic lymphatic leukemia.

continued to be excreted at a markedly elevated level during stimulation of the adrenal gland by ACTH.

Through the cooperation of Dr. Fuller Albright, we were also able to study the urinary steroid excretion of a second breast cancer patient (R. H.) before, during, and after administration of ACTH (Fig. 4). Again a slightly increased ketosteroid excretion was seen during the period of administration of 38 mg ACTH per day. The excretions of the individual steroids are

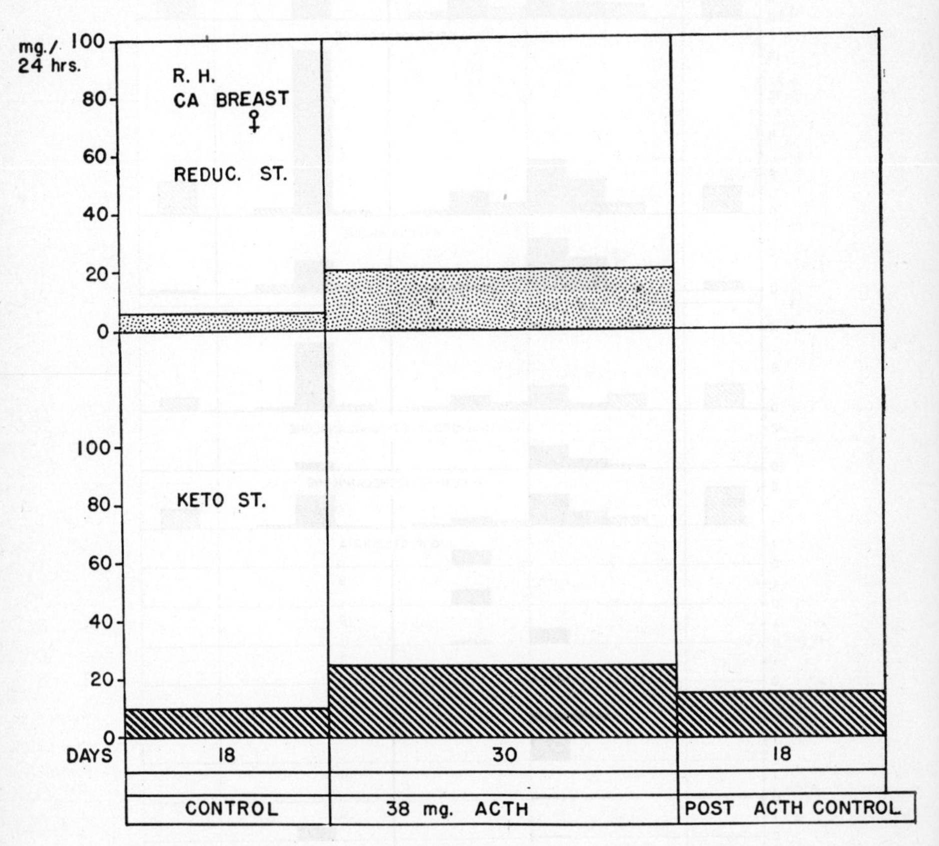

Fig. 4. The excretion of ketosteroids and reducing steroids by a women with cancer of the breast before, during, and after the administration of ACTH.

shown on Fig. 3 (R. H.). All the values were low during the control period but increased significantly during the period of ACTH administration. The excretions of androsterone and etiocholanolone attained the normal range whereas the excretions of the 11-oxygenated steroids, though increased, were still less than normal. In this patient where 11-hydroxyetiocholanolone was absent during the control period, ACTH did not cause the appearance of this compound in the urine.

In a third patient with lymphatic leukemia (I. S.) the steroid excretion before, during, and after administration of ACTH was followed. It can be seen in Fig. 5 that during the administration of 100 mg of ACTH for a period of 6 days no significant change of the steroid excretion occurred. This patient, like the first one which we discussed, was unresponsive to this dose, but showed a greatly increased excretion of steroids when 200 mg of ACTH per day was given. Examination of the excretions of the individual steroids shows that these were extremely low or absent during the control period and

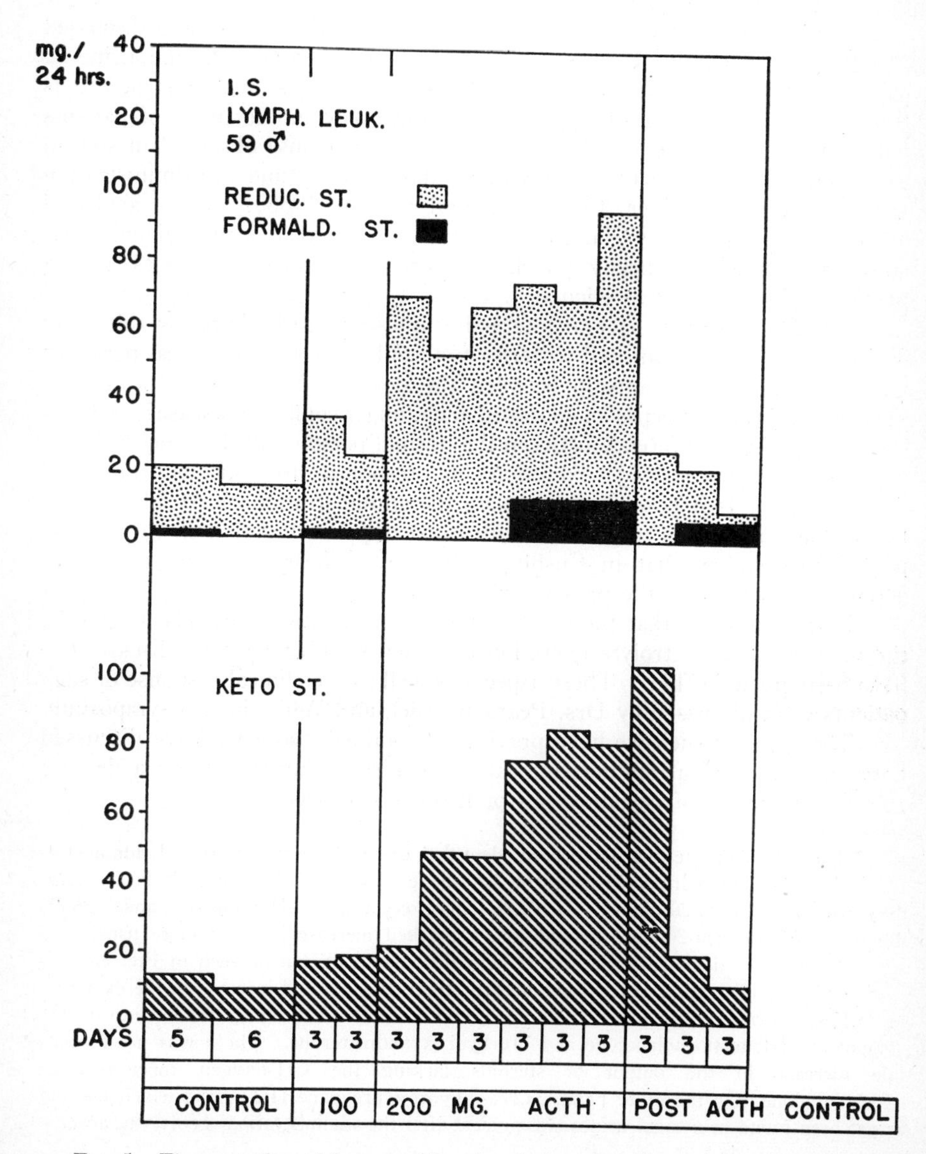

FIG. 5. The excretion of ketosteroids, of reducing steroids, and formaldehydogenic steroids by a man with chronic lymphatic leukemia before, during, and after the administration of ACTH.

that the daily administration of 200 mg of ACTH led to a considerable increase for each compound (Fig. 3) (I. S.). The rise in the excretion of etiocholanolone was most striking being about four times the value for normal healthy subjects and about forty times that found during the control period of this patient. The compound, 11-hydroxyetiocholanolone, which was absent during the control period, appeared in appreciable amounts during the period of ACTH administration.

The administration of ACTH in the three patients we have discussed also led to the excretion of steroids not hitherto observed in the urine of normal persons (Fig. 3).* Some of these compounds, whose structure is thus far unknown, have been recognized previously in the urine of patients with Cushing's syndrome. The type of ACTH response observed in steroid excretion in other patients with various dosage and time of administration varied greatly (Dobriner, Lieberman, and Wilson, 1950). There appeared to be great individual variation when 100 mg of ACTH was given, some subjects showing practically no change whereas in other instances a very pronounced response was elicited.

Our findings raise several interesting questions: (1) What is the effect of ACTH on the amount and pattern of steroid excretion in normal persons? (2) What is the relationship between the dose of ACTH and the steroid excretion in various types of diseases, including neoplastic disease? (3) Is there a significant difference between the normal person and the cancer patient with respect to steroid excretion in response to various doses of ACTH? (4) Is it possible to regulate the administration of ACTH so that the steroid production in patients with cancer approaches that in the normal person rather than that in Cushing's disease? These questions are under active investigation at the present time.

It is of interest that the extent of steroid excretion parallels in general the nitrogen and electrolyte excretion in patients with neoplastic disease who have been given ACTH. These aspects as well as the clinical response of such patients were discussed by Drs. Pearson, Eliel, and White in this symposium.

The results which we have previously obtained, those we have discussed here, and the clinical findings indicate that adrenal function plays a definite role in the course of certain types of neoplastic disease.

---

* The following steroids have been identified in the urine after ACTH administration: $11\beta,17\alpha,21$-trihydroxy-Δ^4-pregnene-3,20-dione (Kendall's Compound F), $3\alpha,17\alpha$-dihydroxypregnane-11,20-dione, $3\alpha,17\alpha,21$-trihydroxypregnane-11,20-dione and $3\beta,21$-dihydroxy-Δ^5-pregnen-20-one. In addition, a marked increase in other more usual constituents such as etiocholanolone and 11-ketoetiocholanolone can be seen in Fig. 3. Following the administration of cortisone acetate, there is a marked increase in the excretion of $3\alpha,17\alpha,21$-trihydroxypregnane-11,20-dione, and a smaller but significant increase in the amount of 11-ketoetiocholanolone and 11β-hydroxyandrosterone. There was no discernible increase in the output of steroids lacking the C11-oxygen function, e.g., etiocholanolone. On the other hand, $3\alpha,17\alpha$-dihydroxypregnane-11,20-dione which has not as yet been found in normal urines was present after the administration of cortisone acetate.

REFERENCES

DOBRINER, K., The excretion of steroids in health and in disease. *Acta union internat. contre cancer,* **6,** 315-328 (1948).

DOBRINER, K., LIEBERMAN, S., and RHOADS, C. P., Methods for the isolation and quantitative estimation of neutral steroids present in human urine. *J. Biol. Chem.,* **172,** 241-261 (1948).

DOBRINER, K., and LIEBERMAN, S., "The Metabolism of Steroid Hormones in Humans," in Gordon, E. S., Editor, *A Symposium on Steroid Hormones.* University of Wisconsin Press, Madison, 1950.

DOBRINER, K., LIEBERMAN, S., and WILSON, H., Adrenal function and steroid excretion in neoplastic disease. *Conference on Use of ACTH.* Armour & Co., 1950.

DOUGHERTY, T. F., and WHITE, A., Effect of pituitary adrenotropic hormone on lymphoid tissue. *Proc. Soc. Exptl. Biol. Med.,* **53,** 132-133 (1943).

HEILMAN, F. R., and KENDALL, E. C., The influence of 11-dehydro-17-hydroxycorticosterone (compound E) on the growth of a malignant tumor in the mouse. *Endocrinology,* **34,** 416-420 (1944).

LIEBERMAN, S., and DOBRINER, K., "Steroid Excretion in Health and Disease. I. Chemical Aspects," in *Recent Progress in Hormone Research,* Vol. III, pages 71-102. Academic Press, New York, 1949.

PEARSON, O. H., ELIEL, L. P., RAWSON, R. W., DOBRINER, K., and RHOADS, C. P., ACTH- and cortisone-induced regression of lymphoid tumors in man. *Cancer,* **2,** 943-945 (1949).

PEARSON, O. H., ELIEL, L. P., and RAWSON, R. W., Regression of lymphoid tumors in man induced by ACTH and cortisone. *Conference on Use of ACTH.* Armour & Co., 1950.

ISOLATION OF ADRENAL CORTICAL HORMONES FROM URINE*

HAROLD L. MASON
Mayo Foundation, Rochester, Minnesota

It has been amply demonstrated that in normal urine there is present material with glycogenic activity similar to that of some of the adrenal cortical hormones which have an oxygen atom at C-11 (Perla and Gottesman, 1931; Weil and Browne, 1939; Dorfman *et al.*, 1942; Horwitt and Dorfman, 1943; Schiller and Dorfman, 1943; Shipley *et al.,* 1943; Venning and Kazmin, 1946). The exact nature of this active material has not been determined. Although some unknown adrenal hormone or a metabolite of some known hormone may be responsible for this activity, it has seemed more likely that one (or more) of the known adrenal cortical hormones is excreted in the urine and is responsible for the observed activity. Several chemical methods for the determination of "corticosteroids" have been based on the likelihood that any compound with the physiologic properties of an adrenal cortical hormone would have the characteristic α-ketol side chain (Talbot *et al.,* 1945; Heard and Sobel, 1946; Corcoran and Page, 1948). Although not entirely specific, the reaction of urinary extracts with periodic acid to give formaldehyde, a property of a primary α-ketol group, has indicated the presence of such "formaldehydogenic" compounds (Corcoran and Page, 1948; Daughaday *et al.,* 1948).

It was found recently that when large amounts of formaldehydogenic material were excreted in a case of Cushing's syndrome, 17-hydroxycorticosterone could be isolated from the urine (Mason and Sprague, 1948) (Fig. 1). This successful isolation of an adrenal cortical hormone from the urine stimulated an attempt to isolate similar compounds from postoperative urine,

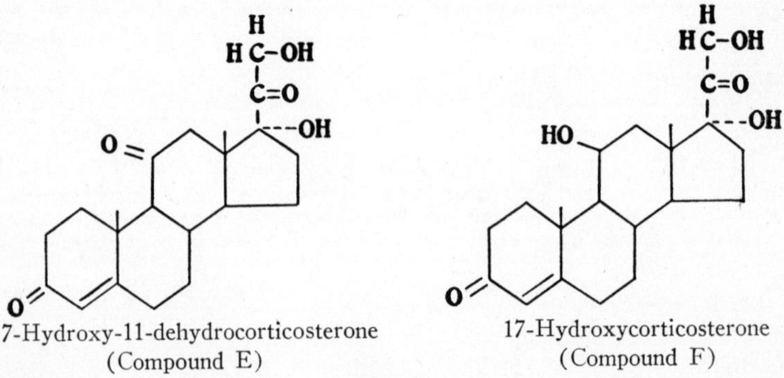

17-Hydroxy-11-dehydrocorticosterone (Compound E)

17-Hydroxycorticosterone (Compound F)

FIG. 1.

* This paper has been previously published in the *Journal of Biological Chemistry,* **182,** 131-149 (1950).

which has been shown to contain an increased amount of biologically active material. Further studies were made on urine obtained from patients suffering from rheumatoid arthritis who were treated with adrenocorticotropic hormone and from patients treated with relatively large amounts of 17-hydroxy-11-dehydrocorticosterone and its acetate (Hench *et al.*, 1949). It is the purpose of this paper to describe the isolation of adrenal cortical hormones from the urine of these patients and from postoperative urine.

Postoperative Urine

In this study* urine was obtained from patients during the five days immediately following operation. The specimens were pooled in the refrigerator with toluene as a preservative and eventually were concentrated in a vacuum and extracted with chloroform. The urine residue was adjusted to *p*H 1 and, after standing for twenty-four hours, it was again extracted with chloroform. These extracts were kept separate at first with the idea of determining whether the acidification had liberated some compound or compounds different from those present in the free state. Actually, the amount of formaldehydogenic substances in the extract obtained after addition of acid was so small in comparison with that obtained before addition of acid that no attempt was made to isolate anything from this small fraction. Later it was combined with one of the major fractions. The toluene used as a preservative extracted very little formaldehydogenic material.

After approximately 100 liters of urine had been extracted the extract was carried through a partial fractionation. In this first attempt at isolation the losses were evidently rather large. With the information thus obtained as a guide, the extract of the second part of the pool obtained from unacidified urine, which contained approximately 110 mg of formaldehydogenic material, was subjected to repeated distributions between benzene and water as described in the experimental part of this paper. The material thus obtained finally in aqueous solution was extracted with chloroform, combined with a similar fraction from the extract of the total urine after acidification to *p*H 1, and was treated in methanol solution with Girard's reagent T (Girard and Sandulesco, 1936) and acetic acid. The Girard complex was decomposed fractionally as described by Reichstein (1936).

The ketonic fractions so obtained were diluted with 95% alcohol to concentrations of 4.7 to 6.0 mg per 100 ml and their absorption in the region 234 to 244 mμ was examined with a Beckman spectrophotometer. The ketonic fraction which was liberated when the solution of the Girard complex was acidified with hydrochloric acid until it turned Congo red paper a pure blue was the only fraction which showed an absorption maximum in this region. From the extinction at 239 mμ, assuming a molecular weight of 360 and a molar extinction coefficient of 16,000 for the material absorbing maximally at this wavelength, it was calculated that 28% of this fraction, or 37 mg was an α,β-unsaturated ketone, probably of the nature of 17-hydroxy-11-dehydro-

* I am indebted to Dr. R. G. Sprague who made the arrangements for collection of urine used in this study and in the study of patients with Addison's disease.

corticosterone. This fraction was combined with two similar fractions obtained from the first half of the urine pool which weighed 26 mg and apparently contained 15 mg of α,β-unsaturated ketone. Attempts to obtain crystals from these combined fractions by treatment with small amounts of chloroform were unsuccessful. Therefore, the solvent was removed and the residue was acetylated with acetic anhydride and pyridine. Crystals were obtained from methanol which gave a green fluorescence with concentrated sulfuric acid, a reaction which is given by 17-hydroxycorticosterone and its acetate. However, after recrystallization the acetate melted at 233-234° and no longer gave the fluorescence with sulfuric acid. It reduced ammoniacal silver promptly in the cold, but did not form an insoluble dinitrophenylhydrazone. These properties indicate the presence of an α-ketol side chain and absence of a 3-keto group.

The mixture of the acetates, which remained in the mother liquors, was subjected to chromatographic analysis on a column of magnesium silicate with mixtures of benzene and petroleum ether, benzene and ethyl ether, and benzene and alcohol. The fraction eluted with benzene-ether, 4/1, proved to be identical with the material which was obtained by direct crystallization from methanol. The next crystalline fraction, which was eluted with benzene-ether, 1/1, melted at 245-250° and gave a green fluorescence with concentrated sulfuric acid. The melting point was depressed about 20° by 17-hydroxy-11-dehydrocorticosterone acetate (m.p. 237-239°). A fraction which weighed 4 mg was eluted with dry ether. After recrystallization from methanol it melted at 220-222°, a mixture with 17-hydroxycorticosterone acetate (m.p. 218-220°) melted at 218-220°, and it gave a strong green fluorescence with sulfuric acid. These properties indicated that this fraction was 17-hydroxycorticosterone acetate, but the small amount of material precluded a more certain identification.

A fourth fraction was eluted with benzene containing 5% of alcohol. After recrystallization from ethanol and methanol it melted at 225-228° and also gave a green fluorescence with concentrated sulfuric acid. This fraction reduced ammoniacal silver promptly in the cold and gave a red dinitrophenylhydrazone. These properties indicated the presence of an α-ketol side chain and an α,β-unsaturated ketone group, as well as an 11-hydroxyl group. Assuming that the structural features indicated are in the positions usual in the adrenal steroids, the minimal structural requirements for these properties would be those of corticosterone, but the acetate of this hormone melts at 152-153°. Reichstein's substance E (Δ^4-pregnene-11(β),17(α),20,21-tetrol-3-one, m.p. 229-230°) was considered as a possibility, but it does not reduce alkaline silver in the cold (Reichstein and Shoppee, 1943). The urinary substance possibly is more highly oxygenated than 17-hydroxycorticosterone since it was removed from the column only when a fairly high concentration of alcohol in benzene was used whereas the fraction with the properties of 17-hydroxycorticosterone was eluted with ether.

Thus the fraction tentatively identified as 17-hydroxycorticosterone acetate was the only fraction with the properties of the acetate of any of the known adrenal cortical steroids.

Patients Treated with ACTH

The urine of 3 women (cases 1, 2, and 3) and 1 man (case 4) with rheumatoid arthritis who had received adrenocorticotropic hormone (ACTH) was made available.* The ACTH† was given in doses of 100 mg per day for twelve days. During this period there was an increase in the formaldehydogenic substances which, toward the end of the period of administration, reached peaks of approximately 12 mg, 17 mg, and 7 mg in the case of the women, and 6 mg per day in the case of the man. Since part of the urine was used for other purposes, extracts were available which in cases 1 and 2 contained approximately 50 mg of formaldehydogenic substances. In the first case the pool of extracts in 10% alcohol was concentrated under reduced pressure with a bath temperature less than 50° in order to remove the alcohol, and the material was distributed between benzene and water five times according to the general procedure described in the experimental part. The benzene was removed from the final fraction and the residue was crystallized from a little chloroform. The crystals weighed 9.5 mg, melted at 208-210°, gave a strong green fluorescence with concentrated sulfuric acid, and immediately reduced ammoniacal silver. After recrystallization from methanol the material melted at 214-216°, and the melting point of a mixture with 17-hydroxycorticosterone was not depressed. Approximately half of the crystals were acetylated with acetic anhydride and pyridine at room temperature. After crystallization from methanol the melting point of the acetate was 214-216°, and it was not depressed by admixture of 17-hydroxycorticosterone acetate.

Fractionation of the material in the chloroform mother liquor with the aid of Girard reagent T and treatment of the ketonic fraction, which was liberated upon addition of sufficient hydrochloric acid to turn Congo red paper blue with acetone and petroleum ether, gave a few more crytsals which had the same properties as those obtained by direct crystallization from chloroform. Acetylation of the noncrystalline material in the mother liquor and crystallization from methanol-water yielded a few crystals of the acetate. The amount of additional crystalline material obtained by these procedures was estimated to be not more than 1 mg.

It was thought that the rather low yield of crystalline material in the first case may have resulted from losses during the distribution between benzene and water and during the later treatment. Consequently the pool of extracts in the second case, after removal of alcohol, was subjected only to two distributions between water and benzene. The final aqueous solution was extracted with chloroform, and the extract was filtered and concentrated at a low temperature in a vacuum to a small volume. Crystals began to separate when the glass was scratched with a spatula. After refrigeration overnight the crystals which were washed with chloroform weighed 24.3 mg.

---

* I am indebted to Drs. P. S. Hench, C. H. Slocumb, and H. F. Polley for their cooperation in making available this urine and that of other patients treated with 17-hydroxy-11-dehydrocorticosterone.

† The ACTH was made available through the courtesy of Dr. John R. Mote, The Armour Laboratories, Chicago, Illinois.

After recrystallization from methanol they melted at 212-215° and a mixture with 17-hydroxycorticosterone (m.p. 214-216°) melted at 212-215°. They gave an intense green fluorescence with concentrated sulfuric acid. $[\alpha]_D^{27} = +168° \pm 5°$ ($c = 0.310\%$). The acetate melted at 217-219°, and the melting point was not depressed by admixture with 17-hydroxycorticosterone acetate which melted at the same point. Since all these properties agreed with the properties of 17-hydroxycorticosterone the identification was considered to be complete. No more crystals could be obtained from the chloroform mother liquor. After removal of chloroform the residue was acetylated, but failed to give crystals of any acetate.

In cases 3 and 4 the procedure was changed somewhat, as detailed in the experimental section. Chloroform solutions of the extract were submitted to fractionation on a column of a 1/1 mixture of magnesium silicate and Celite. In case 3, 6.5 mg was separated from chloroform before the solution was placed on the chromatographic column. Chloroform alone eluted from the column material similar to the nonketonic fraction obtained previously in the Girard fractionation. It melted at 233°, contained nitrogen, was quite soluble in water, and was identified as caffeine. A 3/1 mixture of chloroform and acetone eluted a fraction from which 17-hydroxycorticosterone could be obtained by crystallization from chloroform. In case 3, 6 mg was obtained from this fraction which appeared to contain approximately 15 mg of 17-hydroxycorticosterone as indicated by measurements of optical activity and absorption at 238 mμ. The total of 12.5 mg of 17-hydroxycorticosterone isolated in this case was derived from 37 mg of formaldehydogenic substances in the original pool of extracts.

In case 4, 7.5 mg of 17-hydroxycorticosterone was obtained from 26 mg of formaldehydogenic substances as determined by assay of the pool of extracts.

Patients Receiving 17-Hydroxy-11-dehydrocorticosterone and Its Acetate

The urine of 2 patients with Addison's disease and 4 patients with rheumatoid arthritis who were receiving 17-hydroxy-11-dehydrocorticosterone or its acetate was studied in an effort to identify metabolites of this hormone. Only the isolation of the unchanged hormone will be considered here.

One of the patients with Addison's disease received 100 mg per day of 17-hydroxy-11-dehydrocorticosterone acetate for sixteen days, and the other received 50 mg of the acetate for twenty-three days. Extracts of the urine were made daily during the period of administration of the hormone, and for twenty-seven days after administration of the hormone had been stopped in the first case, since quantitative determinations indicated a continued relatively high excretion of formaldehydogenic substances during this time. All the extracts were combined, and the solvent was removed. The residue was distributed once between benzene and water, and the water-soluble fraction of 687 mg was recovered by extraction with chloroform. It was treated with Girard reagent T, and the Girard complex of the ketonic fraction was subjected to fractional decomposition. In this case also the major ketonic fraction of 56 mg was liberated when the acidity of the solution was adjusted to turn Congo red paper blue. This fraction deposited crystals when the chloroform solution was concentrated to a small volume. Two crops of 3.5 mg each were obtained. The first crop melted at 216-218°. A mixture with 17-hydroxy-11-dehydrocorticosterone (natural, m.p. 218-220°) melted at 216-218°. The crystals reduced ammoniacal silver immediately in the cold and gave a green fluorescence with concentrated sulfuric acid.

The second crop of crystals, which melted at 209-212°, was converted to the acetate with pyridine and acetic anhydride at room temperature. After crystallization from methanol twice, it melted at 238-240° and still gave a green fluorescence with concentrated sulfuric acid. A mixture with 17-hydroxy-11-dehydrocorticosterone acetate (m.p. 240-242°) melted at 238-240°.

The urine of 4 patients, 3 men and 1 woman, with rheumatoid arthritis was collected and extracted daily while 100 mg per day of 17-hydroxy-11-dehydrocorticosterone (Fig. 1) was being administered by intramuscular injection of an aqueous suspension containing 25 mg per milliliter. The urine was collected over a period during which 19.8 g of the hormone was administered. The extracts were pooled in the refrigerator and worked up in two portions. Each portion was distributed several times between benzene and water as described in the experimental part. The aqueous fractions were tested for the presence of α-ketol group by its power to reduce ammoniacal silver in the cold, and for the α,β-unsaturated ketone group by its ability to form a red or orange dinitrophenylhydrazone. These fractions were also assayed for formaldehydogenic substances. The combined aqueous fractions, which gave a strong reduction of ammoniacal silver and a red or orange dinitrophenylhydrazone, assayed 124.6 mg of formaldehydogenic material. The aqueous solution was concentrated and extracted with benzene. The benzene extract was treated with Girard reagent T, and the ketonic fraction was decomposed fractionally. The ketonic fraction liberated, when the solution turned Congo red paper blue, weighed 90 mg and deposited crystals as its chloroform solution was concentrated to a small volume. The crystals weighed 15 mg and melted at 217-219°. The filtrate, after removal of chloroform and treatment with acetone and petroleum ether, gave a second crop of 6 mg which melted at 225-226° after recrystallization from methanol. Both crops gave a strong green fluorescence with concentrated sulfuric acid. When the two crops of crystals were combined and recrystallized from methanol, large prisms were obtained which melted at 227-229°. The fluorescent reaction with sulfuric acid was still strong. This material was identified as 17-hydroxy-11-dehydrocorticosterone by its specific rotation ($[\alpha]^{25}_{D} = +204° \pm 4°$), by observation of an absorption maximum at 240 mμ ($\epsilon = 16{,}200$), and by the melting point of its acetate (240-242°) which was not depressed by admixture of 17-hydroxy-11-dehydrocorticosterone acetate (m.p. 240-242°).

The melting point of highly purified synthetic 17-hydroxy-11-dehydrocorticosterone has been found* to be 230-231° or more than 10° higher than

* Private communication from Dr. V. R. Mattox.

the best melting point reported for the natural compound. As in the case of the compound obtained from adrenal extracts, the compound obtained from urine could not be freed by recrystallization or acetylation from the substance giving the green fluorescence with concentrated sulfuric acid.

COMMENT

The fraction with the properties of 17-hydroxycorticosterone acetate isolated from postoperative urine was disappointingly small. In working with urinary fractions known to contain this hormone in the free state one gains the impression that it is very readily destroyed. It is not likely that it separated quantitatively from chloroform in the presence of much other material; yet exhaustive efforts to obtain more from the mother liquor yielded very little. In the case of the postoperative urine it may be that the preliminary concentration of the urine had an adverse effect.

The failure to find any fraction which could be identified as 17-hydroxy-11-dehydrocorticosterone indicates that this substance was not present in significant amounts since it is more easily isolated than 17-hydroxycorticosterone.

The isolation of 17-hydroxycorticosterone after stimulation of the adrenal cortex with ACTH suggests that this hormone may well be the principal, if not the only, glycogenic steroid which is excreted in the urine normally and under conditions of stress. This suggestion received some support from the isolation of the fraction tentatively identified as 17-hydroxycorticosterone from postoperative urine, although the amount isolated was too small to be very significant.

In the case of 17-hydroxy-11-dehydrocorticosterone isolated from the urine of patients with Addison's disease, the green fluorescence which developed when these crystals were treated with concentrated sulfuric acid indicated the presence of an impurity. This reaction is not specific, but it is given only by compounds with an 11-hydroxyl group. Although early reports (Reichstein, 1936; Wintersteiner and Pfiffner,1936) on preparations of 17-hydroxy-11-dehydrocorticosterone isolated from adrenal extracts indicated that the green fluorescence developed by concentrated sulfuric acid was a property of this substance, Reichstein and Shoppee (1943) pointed out that the completely pure monoacetate does not give the reaction, and it is now evident that the synthetic compound does not give it. Manifestation of the reaction by the crystals isolated from the urine is suggestive of the presence of a small amount of 17-hydroxycorticosterone, which gives the reaction very strongly. It is likely that the presence of this compound was responsible for the reports that 17-hydroxy-11-dehydrocorticosterone isolated from adrenal glands gave this reaction. Several other adrenal steroids also give the reaction and it is possible that one of these may have been responsible in the case of the material isolated from urine. In any event the indications are that the 11-carbonyl group of some of the 17-hydroxy-11-dehydrocorticosterone administered had undergone reduction to an 11-hydroxyl group. Endogenous production of cortical steroids is improbable in view of the severe Addison's disease of long standing which characterized the patients in this study.

In the case of the patients with rheumatoid arthritis who were treated with 17-hydroxy-11-dehydrocorticosterone, the product isolated also gave a green fluorescence with concentrated sulfuric acid even though the melting point approached that of the synthetic hormone. The impurity giving the fluorescent reaction in this instance may have come from the intact adrenals of these arthritic patients, although there was reason to think that their adrenal cortical function was largely inhibited by the administration of 17-hydroxy-11-dehydrocorticosterone.

The study of postoperative urine throws some light on the nature of the glycogenic activity of the urine under conditions of stress, but the small amount of the compound isolated and tentatively identified as 17-hydroxycorticosterone leaves much to be desired. However, when consideration is given also to the isolation of 17-hydroxycorticosterone in satisfactory quantities from the urine of the 4 patients who received ACTH and the 1 patient [previously described (Mason and Sprague, 1948)] with cortical hyperplasia, the importance of this particular hormone in relation to human adrenal function and to the urinary glycogenic activity becomes evident.

It has been assumed that the relief of rheumatoid arthritis and related diseases on administration of ACTH resulted from a greatly increased production of cortisone or a similar substance by the adrenal cortex under the stimulus of ACTH. The isolation of 17-hydroxycorticosterone from the urine of patients treated with ACTH affords strong support for this general assumption although it appears that cortisone may not be produced in significant amounts. Instead, 17-hydroxycorticosterone, a close relative of cortisone, is produced in relatively large amounts and therefore probably is the hormone which is active in the relief of arthritis.

Experimental

Collection and Extraction of Urine. The postoperative urine was collected from 8 to 10 patients at a time and was brought to the laboratory daily where it was pooled in 5-gallon bottles containing toluene and stored in a cold room. When a bottle was filled, the urine was concentrated to a volume between 2 and 3 liters by distillation in a vacuum at a bath temperature not more than 50°. It was necessary to keep the bath temperature less than 50° until the toluene had been removed in order to avoid overheating the urine. After that the urine distilled smoothly while remaining at a relatively low temperature. The concentrate was extracted three times with 500 ml of chloroform. The emulsions that formed were broken by filtration through a pad of infusorial earth. The extract was washed once with a dilute solution of sodium carbonate and then with water until neutral. The extract of each batch was assayed for formaldehydogenic substances which amounted to approximately 0.7 to 1.5 mg per liter of urine. The urine residues were adjusted to approximately *p*H 1 with hydrochloric acid and allowed to stand overnight at room temperature. They were then extracted three times with 500 ml of chloroform. The amount of formaldehydogenic substances varied considerably but averaged approximately 0.3 mg per liter of urine.

In all other cases the urine was received at the laboratory daily and was adjusted to pH 1. It was allowed to stand overnight at room temperature, and then was extracted four times with 0.15 volume of chloroform. The extract was washed as described, and part was used for the determination of formaldehydogenic substances. The remainder was stored in the refrigerator.

Concentration and Distillation. In all cases distillations were conducted in the vacuum produced by an efficient water aspirator. The bath temperature was not allowed to rise above 50°, and when a solution was taken to dryness the temperature was kept under 40°. It is not known with certainty that these precautions are necessary. However, it appears that 17-hydroxycorticosterone may be easily lost. This hormone can be recrystallized readily with the use of heat when it is relatively pure, but in the presence of other urinary constituents heating a chloroform solution makes it very difficult or impossible to obtain crystals even though the hormone is known to be present. Therefore it was deemed advisable to work at relatively low temperatures at all times.

Distribution between Benzene and Water. Although the details varied somewhat from time to time, all the distributions followed the same pattern. One example will be described in detail. The extract of 139 liters of urine, obtained from patients with arthritis who were receiving 17-hydroxy-11-dehydrocorticosterone, was dissolved in 400 ml of benzene. This solution was extracted twelve times with 400 ml of water, leaving "benzene no. 1." The aqueous extract was concentrated to 800 ml and extracted seven times with 400 ml of benzene, leaving "water no. 1." The benzene extract was concentrated to 300 ml and extracted ten times with 300 ml of water, leaving "benzene no. 2." The aqueous extract was concentrated to 400 ml and

TABLE 1

Quantitative and Qualitative Reactions of Fractions Obtained by Benzene-Water Partitions: Extract of Urine from Patients with Arthritis

Fraction	Formaldehydogenic Substances	Reduction of Ammoniacal Silver	Dinitrophenylhydrazone
"Benzene no. 1"	7.7	Weak	None
"Benzene no. 2"	3.4	None	None
"Benzene no. 3"	5.9	None	None
"Water no. 1"	55.8	Weak	Yellow
"Water no. 2"	24.5	Weak	Yellow
Final water	43.3	Strong	Orange

extracted five times with 200 ml of benzene, leaving "water no. 2." The benzene extract was concentrated to 120 ml and extracted with 120 ml of water, leaving "benzene no. 3." The final aqueous solution and the various residues were assayed for formaldehydogenic substances, and qualitative tests were made for the α-ketol group (reduction of ammoniacal silver in the cold) and the α,β-unsaturated ketone group (red or orange dinitrophenylhydrazone). These tests are not highly specific, but they furnished a rough guide as to which fraction or fractions contained material with the properties of an adrenal cortical hormone. The results of this particular fractionation are given in Table 1. It will be noted that the "final water" was the only fraction in which a strong reduction of silver and an orange dinitrophenylhydrazone were combined. These tests indicated that only in this fraction should one expect to find appreciable amounts of an adrenal cortical hormone.

In general, the final aqueous solution was concentrated and extracted three to five times with chloroform. The solvent was removed and the residue was crystallized from chloroform or methanol. If crystallization failed at this stage, fractionation of the Girard complex was tried.

Formation of the Girard Reagent T-Ketone Complex and Its Fractional Decomposition. The residues remaining after removal of organic solvents were dissolved in 5 to 10 ml of methanol, depending on the amount of residue, Girard reagent T was added in an amount double the estimated weight of the residue, and then 2 ml of glacial acetic acid. The mixture was allowed to stand for twenty-four to forty-eight hours at room temperature in a tightly stoppered flask. At first the acetic acid was carefully neutralized when water and ice were added to the mixture and then a small ketonic fraction was obtained when the solution was acidified to litmus. Since this fraction was found to have no value, the procedure was changed slightly. The solution containing the Girard complex, methanol, and acetic acid was poured into a mixture of ice, water, and 6.2 ml of 5*N* sodium hydroxide. The resulting mixture, which was slightly acid to litmus, was extracted three times with chloroform and the extract was considered to be the nonketonic fraction. The aqueous residue was then acidified progressively. A saturated solution of tartaric acid was added until Congo red paper became gray-blue, hydrochloric acid was added until Congo red paper turned pure blue, and finally sufficient hydrochloric acid was added to make its concentration 1*N*. After each adjustment of acidity the solution was allowed to stand at room temperature for one-half to one hour and then extracted with chloroform. Sometimes another fraction was obtained by extraction after standing overnight with *N* hydrochloric acid. Since nothing but amorphous material was obtained from this fraction it will not be considered further. The chloroform extracts were washed free of acid with sodium bicarbonate and water, dried, and distilled.

Melting Points. All melting points were determined with a Fisher-Johns apparatus and are recorded as read. The observed melting points of the compounds here described may vary a few degrees according to conditions. The size of crystals, rate of heating, and temperature at which the sample is placed on the heated block all influence the melting point. For the preparation of mixtures it was necessary to crush the larger crystals, and the reference compound was treated in as nearly as possible the same manner as the unknown. Reference compound, unknown, and mixture were observed on the same cover glass which was placed on the block 15° to 20° below the expected melting point.

Extracts of Postoperative Urine. The extracts of the first 100 liters (first extract) and of the remaining 120 liters (second extract) of urine were worked up separately through the stage of fractionation of the Girard-ketone complex. There were some deviations from the general procedure in the treatment of the first extract and considerable material must have been lost, as indicated in Table 2. These deviations will not be considered further since

TABLE 2

Fractional Decomposition of Keton-Girard Reagent Complex: Extracts of Postoperative Urine

Fraction	First Extract (mg)	α, β-Unsaturated Ketone (mg)	Second Extract (mg)	α, β-Unsaturated Ketone (mg)
1. Nonketone			262	
2. K-1 (Congo gray)	5		12	
3. K-2 (Congo blue)	15	9	133	37
4. K-3 (*N* hydrochloric acid)	11	8	94	
5. K-4 (*N* hydrochloric acid after 24 hours)			15	

they led to poor results. Both extracts were subjected to the benzene-water partition and to the fractionation of the Girard complex. The total extract (containing 63 mg of formaldehydogenic substances) of the urine after acidification to *p*H 1 was carried through the benzene-water partitions and then combined with the "second extract" for the Girard fractionation. Table 2 gives the results of the fractionation of the Girard complexes. The fractions labeled K-2 showed an absorption maximum at 239 mμ, a region in which α,β-unsaturated ketones show an absorption maximum. In the case of the "first extract" the ketonic fraction liberated with *N* hydrochloric acid (K-3) also showed an absorption maximum. These examinations were made with a Beckman spectrophotometer. The concentrations were 1.67, 1.10, and 5.3 mg per cent in 95% alcohol for K-2 and K-3, "first extract" and K-2, "second extract," respectively; $E^{1\%}_{1\,cm} = 268$, 289, and 125, respectively. Assuming the presence of an adrenal cortical hormone with a molecular weight of 360 and a molar extinction of 16,000 the values given in Table 2 for "α,β-unsaturated ketone" were calculated.

These three ketonic fractions were combined and diluted with water. The solution was partially distilled to remove alcohol and extracted with chloroform. Since concentration of the chloroform solution failed to yield crystalline material, the chloroform was removed completely and the residue was dissolved in 2 ml of pyridine and 1 ml of acetic anhydride and allowed to stand overnight at room temperature. Dilute hydrochloric acid and ice were added, and the mixture was extracted with chloroform. After the solution had been washed with sodium carbonate and dried over sodium sulfate, it was filtered and the chloroform was distilled. The residue was taken up in 0.5 ml of methanol from which crystals separated on refrigeration overnight. They melted at 232-233° and a mixture with 17-hydroxy-11-dehydrocorticosterone (m.p. 242-243°) melted below 220°. Combination with a second crop and recrystallization gave a product which melted at 233-234°. The purified material gave no color with concentrated sulfuric acid.

It reduced ammoniacal silver promptly in the cold but did not form an insoluble dinitrophenylhydrazone. The material in the mother liquor (about 200 mg) was chromatographed on a column of 10 g of a 1/1 mixture of magnesium silicate and infusorial earth. Fractions were eluted with mixtures of petroleum ether and benzene, benzene, mixtures of benzene and dry ether, dry ether, and with benzene containing small concentrations of alcohol. Benzene-ether (4/1) removed 21 mg of crystalline material which proved to be identical with the acetate which crystallized directly from the methanol solution of the crude mixture and melted at 233-234°. A second fraction of 16 mg was eluted with benzene-ether (1/1). It melted at 245-250° after crystallization from methanol and gave a green fluorescence with concentrated sulfuric acid. A mixture with 17-hydroxy-11-dehydrocorticosterone acetate (m.p. 237-239°) melted at 225-230°. Dry ether eluted 4 mg of material which melted at 220-222° after recrystallization from methanol and which gave a green fluorescence with sulfuric acid. A mixture with 17-hydroxycorticosterone acetate (m.p. 218-220°) melted at 218-220°. A fourth fraction, removed by 5% by volume of alcohol in benzene, weighed 39 mg and melted at 225-228° after recrystallization from methanol. It also gave the green fluorescence with sulfuric acid, reduced ammoniacal silver promptly in the cold, and formed a red dinitrophenylhydrazone.

Isolation of 17-Hydroxycorticosterone from the Urine of 4 Patients Receiving Adrenocorticotropic Hormone. Part of the urine was used for various determinations and part of the extracts for determination of formaldehydogenic substances. The extracts that remained were pooled in the refrigerator in 10% alcoholic solution. In cases 1 and 2 the pool of extracts assayed approximately 50 mg of formaldehydogenic substances; in case 3, 37 mg; and in case 4, 26 mg (calculated as 17-hydroxycorticosterone). In the first case the alcohol was removed in a vacuum, and the material was subjected to distribution between benzene and water five times. Instead of extracting the final aqueous solution with chloroform it was extracted seven times with an equal volume of benzene. After removal of the benzene, the residue was transferred to a small test tube with chloroform, and this solution was concentrated to about 0.5 ml by warming in an air stream. The crystals which separated on refrigeration overnight were collected on a filter and washed free of color with chloroform. They weighed 9.5 mg, melted at 208-210°, gave a strong green fluorescence with sulfuric acid, and reduced ammoniacal silver immediately in the cold. After recrystallization from methanol the melting point was 214-216°, and a mixture with 17-hydroxycorticosterone (m.p. 214-216°) melted at 214-216°.

Approximately half of the crude material was acetylated overnight at room temperature with 4 drops of pyridine and 2 drops of acetic anhydride. Water and hydrochloric acid were added, and the crude acetate was collected on a filter and washed with hydrochloric acid and water. After recrystallization from methanol, the acetate melted at 214-216° and a mixture with 17-hydroxycorticosterone acetate (m.p. 216-218°) melted at 214-216°.

The material in the mother liquor from the crude 17-hydroxycorticosterone was treated with Girard reagent, and the ketone complex was decomposed fractionally. In this manner 22 mg of nonketone, 5 mg of fraction K-1 (Table 2), 14 mg of fraction K-2, and 6 mg of fraction K-3 were obtained. Fraction K-2 appeared to become partly crystalline on addition of a few drops of chloroform. However, crystallization from methanol failed. Probably it was a mistake not to crystallize from a little chloroform first, since this is the best solvent yet encountered for separation of 17-hydroxycorticosterone from other urinary constituents. A few more crystals of 17-hydroxycorticosterone were obtained after removal of the methanol, by dissolving the residue in acetone and adding petroleum ether slowly while warming. Acetylation of the remaining noncrystalline material and cautious addition of water to a methanol solution of the product gave a few crystals.

Because of what seemed to be a poor yield of 17-hydroxycorticosterone in the first case, the extract in case 2 was not subjected to as extensive a fractionation. After removal of alcohol, the aqueous solution was diluted to 200 ml and extracted fifteen times with 200 ml of benzene. The benzene extract was concentrated to 150 ml and extracted twelve times with 150 ml of water. The aqueous extract was concentrated to 350 ml and extracted four times with 100 ml of chloroform. The chloroform solution was filtered, concentrated to about 3 ml, transferred to a small test tube, and concentrated by warming in an air stream to about 0.5 ml. On scratching the tube crystals began to separate, and after refrigeration overnight they were collected on a filter, washed free of color with chloroform and dried briefly at 115°. They weighed 24.3 mg, melted at 207-208° and had the same other properties as the crystals obtained in the first case. After recrystallization from methanol they melted at 212-215°, and a mixture with 17-hydroxycorticosterone (m.p. 214-216°) melted at 212-215°; $[\alpha]_D^{27} = +168° \pm 5°$ ($c = 0.310\%$ in alcohol). The specific rotation of 17-hydroxycorticosterone has been given as $[\alpha]_D^{22} = +167.2° \pm 2°$ and $[\alpha]_D^{26} = +167° \pm 3°$ (Mason and Sprague, 1948; Reichstein, 1937). The acetate melted at 217-219° and a mixture with 17-hydroxycorticosterone acetate (m.p. 217-219°) melted at 217-219°.

The material in the chloroform mother liquor was acetylated but further crystalline material could not be obtained.

In the third and fourth cases the pools of extracts were treated essentially as in the second case. In the third case the chloroform extract so obtained was concentrated to a small volume and 6.5 mg of crystals (m.p. 203-206°) were separated. The filtrate was diluted to 10 ml and poured onto a column of 4 g of a 1/1 mixture of magnesium silicate and Celite. The column was washed with a total of 250 ml of chloroform, which removed crystalline material. This material melted at 233° after recrystallization from methanol. It was identified as caffeine. A 3/1 mixture by volume of chloroform and acetone (200 ml) removed 26 mg of an oily fraction which was dissolved in 10 ml of 95% alcohol and filtered to remove a small amount of insoluble material. The optical rotation of this solution for sodium light was +0.267°. This rotation corresponded to 14.8 mg of 17-hydroxycorticosterone ($[\alpha]_D^{27} = +167°$). The solution was diluted (0.3 ml diluted to 50 ml) and, with the use of a 1-cm cell in a Beckman spectrophotometer, an absorption maximum was observed at 238 mμ with an extinction of 0.402. This extinction corresponded to 15.3 mg of 17-hydroxycorticosterone ($\epsilon = 15{,}800$, mol. wt. = 362) in the undiluted solution. The substantial agreement of the weights calculated from the optical rotation and from the extinction at 238 mμ suggests that 17-hydroxycorticosterone was the only substance present which affected these measurements. After removal of alcohol the fraction yielded 6 mg of crystals (m.p. 203-205°) from a small volume of chloroform. These crystals were combined with those obtained previously and recrystallized from methanol. The melting point, 212-214°, of the purified product was not depressed by admixture of 17-hydroxycorticosterone.

In the last case the pool (120 ml) of extracts assayed 26 mg of formaldehydogenic substances calculated as 17-hydroxycorticosterone. Ten extractions with 100 ml of benzene removed 19.6 mg. The benzene solution was concentrated to 100 ml and extracted ten times with 100 ml of water. The combined aqueous extracts assayed 20.5 mg, indicating that all the formaldehydogenic substances had gone from the benzene solution into the aqueous extracts. The aqueous solution was concentrated to 200 ml and extracted four times with 50 ml of chloroform. This extract was dried over sodium sulfate and concentrated to 50 ml. It was then passed through a column of 3 g of the magnesium silicate-Celite mixture. The pattern of elution was similar to that described in case 3. Chloroform-acetone (3/1) removed 28 mg of material from which 7.5 mg of crystals were obtained on crystallization from 0.5 ml chloroform. The crystals melted at 209-211°. After recrystallization from methanol they melted at 214-216° and a mixture with 17-hydroxycorticosterone melted at 214-216°.

The preparations obtained in the last two cases were combined and 4.697 mg were dissolved in 3 ml of 95% alcohol. $[\alpha]^{27}_{D} = 166° \pm 6°$. The alcohol was evaporated by warming in a stream of air, and the residue was acetylated overnight at room temperature. After crystallization from methanol the produce melted at 216-218° and a mixture with 17-hydroxycorticosterone acetate (m.p. 217-219°) melted at 216-218°.

Isolation of 17-Hydroxy-11-dehydrocorticosterone from the Urine of Patients Receiving This Hormone and Its Acetate. Two patients with Addison's disease received a total of 2.75 g of 17-hydroxy-11-dehydrocorticosterone acetate. Daily extracts were made for quantitation of the formaldehydogenic substances and that portion (about 80%) of each extract not required for this purpose was pooled in chloroform solution with others in the refrigerator. Extracts from the urine of both patients were combined. When the collection was complete the chloroform was removed and the residue was dissolved in 100 ml of benzene. This solution was extracted ten times with 100 ml of water and the water solution was extracted five times with 150 ml of chloroform. Distillation of the benzene remaining after extraction with water and of the chloroform gave respectively 275 mg and 687 mg of solids. The 687 mg of material were treated with Girard reagent, and the ketone complex was decomposed fractionally. The nonketonic fraction weighed 498 mg and the ketonic fraction liberated on addition of hydrochloric acid sufficient to turn Congo red paper blue weighed 56 mg. The other ketonic fractions weighed only 10 and 20 mg. The major ketonic fraction (K-2) crystallized on addition of a few drops of chloroform. Two crops of 3.5 mg each were obtained by crystallization from chloroform. The first crop gave a green fluorescence with sulfuric acid and an orange dinitrophenylhydrazone, reduced ammoniacal silver immediately in the cold, and melted at 216-218°. A mixture with 17-hydroxycorticosterone melted at 205-208° whereas a mixture with 17-hydroxy-11-dehydrocorticosterone (m.p. 218-220°) melted at 216-218°. The second crop (m.p. 209-212°) was converted to the acetate which, after recrystallization twice from methanol, melted at 237-240°. A mixture with 17-hydroxy-11-dehydrocorticosterone acetate (m.p. 238-240°) melted at 238-240°.

The remainder (49 mg) of fraction K-2 was acetylated and, since only a trace of crystalline material could be obtained from solvents, the product was chromatographed on a column of magnesium silicate-Celite (1/1). The same sequence of solvents was used as in the case of postoperative urine. Dry ether removed a small fraction which weighed about 1 mg after recrystallization from methanol and which melted at 235-238°. A mixture with 17-hydroxy-11-dehydrocorticosterone acetate melted at 236-239°. Another fraction was removed with 2% by volume of ethanol in benzene. It crystallized from aqueous acetone along with flocks of yellow pigment. The latter was removed by washing with methanol. The crystals melted at 244-246°. They depressed the melting point of 17-hydroxy-11-dehydrocorticosterone acetate and also the acetate melting at 245-247° obtained from postoperative urine.

The material (275 mg) which, after extraction with water, remained in the benzene solution of the orginal combination of extracts, was fractionated with the aid of Girard reagent T. This treatment yielded 234 mg of nonketones and a total of 20 mg of ketones. None of the ketonic fractions showed specific absorption in the neighborhood of 238-240 mμ and none of the fractions, including the nonketonic, gave a detectable amount of formaldehyde on oxidation with periodic acid.

Urine was collected in varying amounts from 4 patients with rheumatoid arthritis who received 100 mg per day of free 17-hydroxy-11-dehydrocorticosterone. Approximately 210 liters of urine were obtained while the patients received 19.8 g of the hormone during 204 patient-days. Treatment of the urine and extraction were the same as in the cases of Addison's disease. The chloroform extracts were processed in two portions.

The results of the benzene-water partitions of the first portion are given in Table 1. Similar treatment of the second portion resulted again in concentration in the final water solution of the material which reduced ammoniacal silver immediately in the cold and which gave an orange dinitrophenylhydrazone. This fraction assayed 81.3 mg of formaldehydogenic material. This aqueous solution was combined with that obtained from the first portion (total formaldehydogenic substances of both portions 124.6 mg), concentrated to 200 ml, and extracted seven times with 100 ml of benzene. The solvent was removed, and the residue was subjected to fractionation with the aid of Girard reagent. The ketonic fraction (K-2) liberated by acidity sufficient to turn Congo red paper blue weighed 90 mg and crystallized on concentration of the chloroform solution to a small volume. The chloroform was removed completely, and the residue was dissolved in methanol. The solution was filtered from a small insoluble fraction, transferred to a small tube and concentrated in an air stream while warming in a bath at 50° to about 1 ml. Crystals appeared only after scratching the walls of the tube. After refrigeration overnight the crystals were collected on a filter, washed with methanol and dried at 115°. They weighed 15 mg, melted at 217-219°, and gave a green fluorescence with concentrated sulfuric acid. The methanol was removed from the filtrate and the residue was dissolved in a little acetone. Cautious addition of petroleum ether led to separation of crystals accompanied by some oil. The oil was removed by washing with 3 drops of methanol, and the remainder was recrystallized from methanol to give 6 mg of material which melted at 225-226°. The two portions of crystalline material (21 mg) were combined and recrystallized from methanol. This product melted at 227-229° when placed on the heated block at 215°. It still gave quite a strong green fluorescence with sulfuric acid. $[\alpha]_D^{25} = +204° \pm 4°$ ($c = 0.250\%$ in alcohol). With a solution of 1 mg per cent in 95% alcohol an absorption maximum was observed at 240 mμ ($\epsilon = 16{,}200$). Acetylation of 7 mg with 5 drops of pyridine and 4 drops of acetic anhydride at room temperature overnight followed by crystallization of the product from methanol gave an acetate which melted at 240-242° and showed a green fluorescence with sulfuric acid. A mixture with 17-hydroxy-11-dehydrocorticosterone acetate (m.p. 240-242°) melted at 240-242°.

Summary

A substance isolated as an acetate which had the properties of 17-hydroxycorticosterone acetate was obtained from the urine of patients who had undergone major surgical operations. The amount was too small to permit certain identification. 17-Hydroxycorticosterone was isolated in significant amounts from the urine of 4 patients suffering from rheumatoid arthritis who received adrenocorticotropic hormone in doses of 100 mg per day. The urine of similar patients who received 100 mg per day of 17-hydroxy-11-dehydrocorticosterone yielded small amounts of this hormone unchanged. Small amounts of 17-hydroxy-11-dehydrocorticosterone also were isolated from the urine of 2 patients suffering from Addison's disease who were treated with the acetate of this substance. There was some evidence of a physiologic reduction of the 11-keto group of 17-hydroxy-11-dehydrocorticosterone to a hydroxyl group. Methods are described for isolation of these adrenal cortical hormones from urine.

Schneider (1950) recently has isolated 17-hydroxy-11-dehydrocorticosterone (cortisone) from the urine of normal men.

REFERENCES

Corcoran, A. C., and Page, I. H., Methods for the chemical determination of corticosteroids in urine and plasma. *J. Lab. Clin. Med.,* **33** (Pt. 2), 1326-1333 (1948).

Daughaday, W. H., Jaffe, Herbert, and Williams, R. H., Chemical assay for "cortin." Determination of formaldehyde liberated on oxidation with periodic acid. *J. Clin. Endocrinol.,* **8,** 166-174 (1948).

DORFMAN, R. I., HORWITT, B. N., and FISH, W. R., The presence of cortin-like substance (cold protecting material) in the urine of normal men. *Science,* **96,** 496-497 (1942).

GIRARD, ANDRÉ, and SANDULESCO, GEORGES, Sur une nouvelle série de réactifs du groupe carbonyle, leur utilisation à l'extraction des substances cétoniques et a la caracterisation microchimique des aldehydes et cétones. *Helv. Chim. Acta,* **19,** 1095-1107 (1936).

HEARD, R. D. H., and SOBEL, H., Steroids. VIII. A colorimetric method for the estimation of reducing steroids. *J. Biol. Chem.,* **165,** 687-698 (1946).

HENCH, P. S., KENDALL, E. C., SLOCUMB, C. H., and POLLEY, H. F., The effect of a hormone of the adrenal cortex (17-hydroxy-11-dehydrocorticosterone: compound E) and of pituitary adrenocorticotropic hormone on rheumatoid arthritis. *Proc. Staff Meetings Mayo Clinic,* **24,** 181-197 (1949).

HORWITT, B. N., and DORFMAN, R. I., The effect of urinary cortin-like material on carbohydrate metabolism. *Science,* **97,** 337 (1943).

MASON, H. L., and SPRAGUE, R. G., Isolation of 17-hydroxycorticosterone from the urine in a case of Cushing's syndrome associated with severe diabetes mellitus. *J. Biol. Chem.,* **175,** 451-456 (1948).

PERLA, D., and GOTTESMAN, J. M., Substance in urine of normal human adults that raises resistance of suprarenalectomized rats. *Proc. Soc. Exptl. Biol. Med.,* **28,** 1024-1027 (1931).

REICHSTEIN, T., Über Bestandteile der Nebennieren-Rinde. VI. Trennungsmethoden, sowie Isolierung der Substanzen F.a. H und J. *Helv. Chim. Acta,* **19,** 1107-1126 (1936).

REICHSTEIN, T., Über Bestandteile der Nebennierenrinde. X. Zur Kenntnis des Corticosterons. *Helv. Chim. Acta,* **20,** 953-969 (1937).

REICHSTEIN, T., and SHOPPEE, C. W., "The Hormones of the Adrenal Cortex," in Harris, R. S., and Thimann, K. V., *Vitamins and Hormones; Advances in Research and Applications,* Vol. 1, pp. 345-413. Academic Press, New York, 1943.

SCHILLER, SARA, and DORFMAN, R. I., Influence of urinary cortin-like material on water intoxication in the adrenalectomized rat. *Endocrinology,* **33,** 402-404 (1943).

SCHNEIDER, J. J., Studies on the excretion of adrenocortical compounds. I. Isolation of 17-hydroxy-11-dehydrocorticosterone and other compounds from urine of normal males. *J. Biol. Chem.,* **183,** 365-376 (1950).

SHIPLEY, R. A., DORFMAN, R. I., and HORWITT, B. N., Presence in normal urine of cortin-like material which is active in muscle work test. *Am. J. Physiol.,* **139,** 742-744 (1943).

TALBOT, N. B., SALTZMAN, A. H., WIXOM, R. L., and WOLFE, J. K., Colorimetric assay of urinary corticosteroid-like substances. *J. Biol. Chem.,* **160,** 535-546 (1945).

VENNING, E. H., and KAZMIN, V., Excretion of urinary corticoids and 17-ketosteroids in normal individual. *Endocrinology,* **39,** 131-139 (1946).

WEIL, P., and BROWNE, J. S. L., Excretion of cortin after surgical operation. *Science,* **90,** 445-446 (1939).

WINTERSTEINER, OSKAR, and PFIFFNER, J. J., Chemical studies on the adrenal cortex. III. Isolation of two new physiologically inactive compounds. *J. Biol. Chem.,* **116,** 291-305 (1936).

ADRENAL CORTEX IN HYPERTENSION

A. C. CORCORAN

Cleveland Clinic Foundation and the Frank E. Bunts Educational Institute, Cleveland, Ohio

Evidences of adrenal participation in hypertension are experimental and clinical. Experimentally, hypertensive vascular disease can be elicited in rats, under suitable conditions by treatment with desoxycorticosterone (DC) and by treatment with crude suspensions of anterior pituitary (LAP or APP). The hypertension in these experiments would seem in the first instance attributable to hypercorticoidism, but, as we shall see, this attribution does not seem to us firmly established.

The best-established relationship is not one of pathogenesis. In renal hypertension adrenalectomy is firmly established as a procedure which causes a decrease in arterial pressure and experiments with hypophysectomized rats treated with ACTH establish this effect as one of hypocorticoidism. It was our impression from earlier experiments with Dr. Irvine H. Page that adequate replacement therapy did not fully restore the hypertension of adrenalectomized renal hypertensive dogs. More recent studies by Drs. Lena A. Lewis and Page (1948) have established that hypertension can be fully restored by treatment with relatively small doses of adrenal cortical extract and/or DC. The difference lies in their skillful attention to operative and postoperative care, minimizing trauma, and avoiding infection. The acclerated post-traumatic utilization of cortical hormones being held in check, the influence of the hormones on arterial pressure becomes fully manifest. It is unequivocally shown that the normal function of the adrenal cortex is essential to the maintenance of a hypertensive state in which it is not genetically concerned. The experiments with DC and LAP cannot be as directly interpreted, but they do indicate that abnormal function of the adrenal cortex may be a cause of hypertensive vascular disease.

Clinically, the severity of nephrosclerosis and hypertensive vascular disease in Cushing's syndrome seems to me to have been underemphasized by the endocrinologists, who, quite naturally, are more fascinated by its unusual metabolic deviations. However, to one primarily interested in vascular disease, the renal vascular consequences of Cushing's syndrome are very striking (Corcoran, Taylor, and Page, 1948). From the aspect of relating the adrenal and hypertension, this syndrome is a sterling experiment by Nature in which hypercorticoidism commonly, if not inevitably, results in severe vascular damage. The hypertension of some cases of phaeochromocytoma may have an adrenal cortical as well as a medullary component. At least, liberation of epinephrine from the tumor should elicit the release of ACTH which should in turn result in cortical hyperfunction. In some of these tumors, the pressor component is *l*-arterenol rather than *l*-epinephrine. *l*-Arterenol does not stimulate release of ACTH. Thus, speculatively, it might be possible to estimate the methylating function of a mass of abnormal adrenal medullary tissue from the results of a survey of adrenal cortical function. But these, and a few other rare conditions, compass the clinical hypertensions of endocrine origin, and all of them lumped together form only a small fraction of the mass of hypertensive vascular disease with which the clinician is called upon to deal.

Workers in the vascular field, entranced by the possibility that the adrenal cortex might be of genetic as well as supportive concern in essential hypertension (hypertension of unascribable origin) are therefore faced, not with a problem to solve, but with the need for orientating studies which will show whether or not the problem exists. With this in mind, I beg your indulgence, if I present, not a single definitive study, but a series of loosely correlated observations directed toward the evalution of adrenal participation in hypertensive vascular disease. Having reviewed them, and seen their concurrences, we may be inclined to agree that a problem does exist, even if the crucial experiments have not been done.

Experimental

A problem of first concern was that of DC hypertension, the very existence of which has at times been called in question, and of hypertension due to LAP. With Dr. Georges Masson, we have shown that DC hypertension in rats is a most regular phenomenon when the conditions of Selye's experiments are duplicated (uninephrectomy, administration of 1% NaCl to drink). The course of the hypertension is predictable. Hypertension is present in all the animals at the end of 3 or 4 weeks; it persists and becomes more severe during the following 4 to 6 weeks. Even the sequences of morphological change in kidneys and other vascular areas can be predicted with some accuracy. Quite apart from its theoretical implications, DC hypertension provides an invaluable tool to the experimentalist.

The hypertension due to injection of crude anterior pituitary powder (LAP) is much more variable. In some animals there is no definite increase of arterial pressure, even when the sensitizing conditions are rigidly observed (uninephrectomy, saline, high protein diet). Some of the irregularity is probably due to formation of antibodies (antihormones?) since administration of gradually incrementing doses of LAP or of alkaline extract of LAP tend to stabilize the curves of blood pressure increase. In passing, we would emphasize that infection and abscess formation at the sites of injection do not explain LAP hypertension or arterial disease, since sterile extracts of LAP elicit the disease but not the abscess. Having thus confirmed in some detail the work of Selye, we turned to a consideration of mechanism (Selye, 1946, 1947).

Selye's working hypothesis was that DC and LAP hypertensions were genetically similar, that injection of LAP causes release of DC-like mineralocorticoids which cause a hypercorticoidism similar to that produced by injection of DC. Direct measurements of DC-like substances which might confirm this have not been feasible. Rather, the view was supported in great measure by the fact that the adrenals of rats treated with LAP are unusually large.

A characteristic of overdosage with DC is polyuria. This is believed to be a homeostatic adjustment to sodium retention. Excretion of a large volume of urine tends to excrete sodium, even when the sodium concentration of the urine is depressed by increased reabsorption due to DC stimulation of tubular cells. Granting this, the occurrence of polyuria during treatment with LAP would tend to confirm the view that there is indeed a hypersection of mineralocorticoids. We have therefore compared (Masson, Corcoran, and Page, 1949) the rates of urine flow in rats treated with DC, with LAP and in untreated controls under identical experimental conditions (Fig. 1). The DC rats showed severe polyuria. But the urine flow of rats treated with

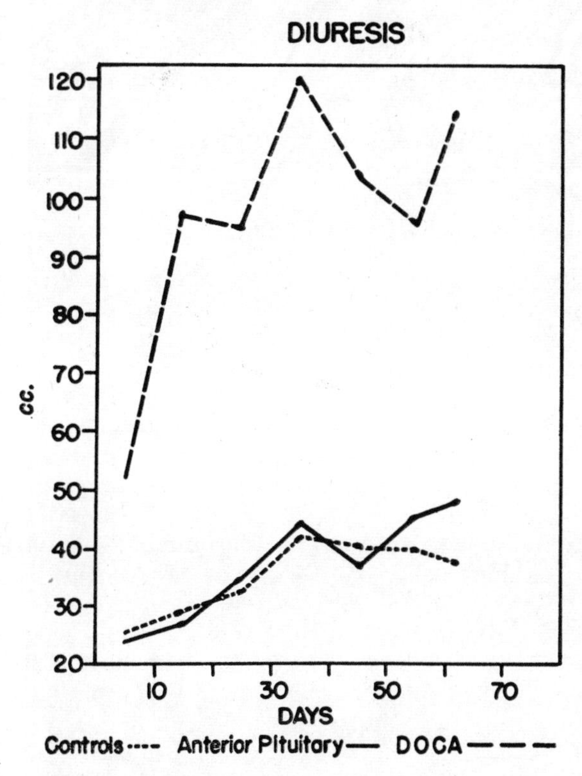

FIG. 1. Urine volume (ordinate) by days of treatment (abscissa) of uninephrectomized rats given 1% NaCl to drink. Comparison of controls, of DCA treatment, and of treatment with lyophilized anterior pituitary.

LAP did not exceed that of the controls. From this we adduce that LAP hypertensive disease is not due to overproduction of mineralocorticoids with consequent sodium retention or that, if it is, the overproduction is somehow very closely balanced by some antidiuretic state. Among the facts supporting the former view, three may be noted: (1) LAP (lyophylized beef anterior pituitary, Desbergers, Montreal) is poor in corticotropin; (2) in any case, corticotropin probably does not have in the rat a direct influence on formation of mineralocorticoids, since it does not affect the zona glomerulosa, and (3)

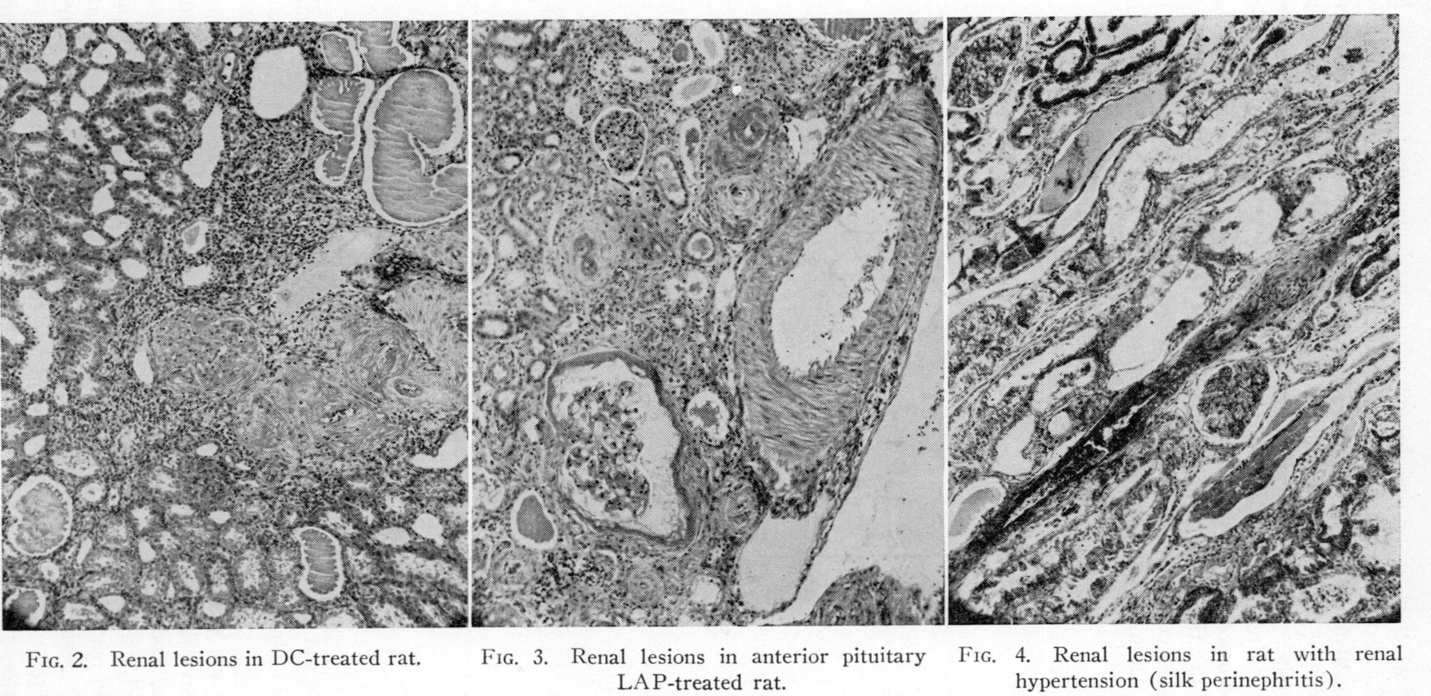

FIG. 2. Renal lesions in DC-treated rat.

FIG. 3. Renal lesions in anterior pituitary LAP-treated rat.

FIG. 4. Renal lesions in rat with renal hypertension (silk perinephritis).

the adrenals of LAP-treated rats, while large, are also often necrotic and hemorrhagic. A study of adrenal morphology in the various forms of hypertension in rats is in progress (Drs. Deane and Masson), and their report will supplement this comment in detail.

Thus, the genetic mechanisms of DC and LAP hypertensive disease seem to us as distinct as are, for example, DC hypertension and renal hypertension. We would question the view that LAP hypertension is attributable to hypercorticoidism and, were we presently so minded, could find support for the thesis that there may be an element of adrenal cortical insufficiency, if not in the hypertension, at least in the vascular disease provoked by LAP. Indeed, one could argue from the small adrenals of rats treated with DC that here again there may be lack or imbalance of certain steroids. Such a point of view would accord well with the views expressed here by Dr. Dougherty.

In taking the view that LAP hypertension is not attributable to hypercorticoidism, we do not imply that adrenal cortical hyperfunction is not a cause of hypertensive disease. Supplementing the fact of Cushing's syndrome, a personal communication from Dr. Selye informs us that he has produced hypertensive vascular disease in rats by injections of purified ACTH. This does not establish that LAP hypertension is due to ACTH. It does show that ACTH must be added to DC, LAP, and kidney wrapping as a cause of hypertension in the rat.

This being the case, at least three, possibly four, genetically distinct forms of hypertensive vascular disease are available for study in rats. With Drs. Masson, Hazard, and Page, we have carefully reviewed (1950) the nature and sequence of morphological changes in rats made hypertensive by DC, by LAP, and by establishment of perinephritis. We are struck in the first place by the uniform character of the lesions provoked by these different causes. The renal lesions are identical with each other and with the renal lesions of malignant nephrosclerosis in human beings. The extrarenal lesions differ from each other only in detail, and are especially severe in the splanchnic arteries. These latter have been described as "periarteritis nodosa," which is truly descriptive, since the exudate invades the vessel walls from without in, but which has an unjustified clinical connotation since it implies an identification with clinical polyarteritis nodosa. For example, renal arteritis is very rare in any of the experimental groups, whereas in clinical periarteritis, it is very common. Consequent, since the splanchnic lesions of rats appear in three different forms of hypertension, we regard them not as equivalents of human periarteritis, but as manifestations of hypertensive arteriopathy specific to the rat.

To review this discussion of experiments, we do not view DC and LAP hypertensions as basically of similar mechanism. Rather, we believe them to be distinct entities. Nor is it proved that either or both are consequences of hypercorticoidism as such. DC may not be a hormone active in nature. Some aspects of both hypertensions are not inconsistent with relative specific deficiencies of certain corticoids, notably those of the zona fasciculata. Both conditions, however, implicate the adrenal in the genesis of hypertensive vascular disease. Morphologically, their manifestations and those of experimental renal hypertension are identical and may be considered manifestations of hypertensive arteriopathy in the rat. Renal lesions in rats under these varied stimuli are identical with those of malignant hypertension in human beings.

Compound S (17-hydroxy-desoxycorticosterone) is present in normal adrenals. It was known to have some activity in electrolyte metabolism but not to influence various aspects of carbohydrate metabolism. Hence, as it became available to us, a further study was planned with Dr. Masson.

TABLE 1

Effects of DCA and Compound S on Blood Pressure

	Blood Pressure mm Hg			
Treatment	8 days	14 days	22 days	31 days
O	125	129	140	140
DCA	149	167	187	194
Compound S	118	124	142	137
DCA-Compound S	146	150	189	200

In the first series of experiments, 30-mg pellet implants were compared with implants of DC as regards diuresis and hypertension under conditions which favor the onset of DC hypertensive disease. In addition, the effects of simultaneous implants of DC and compound S in the same animals were observed for evidences of activity enhancing or inhibiting the actions of DC. The data (Table 1) establish that, under these conditions, compound S does not elicit either polyuria or hypertension and that it has no obvious inhibitory or enhancing effect on the hypertension or the polyuria produced by implants of DC. In a further brief study (10 days) injections of compound S in a dose of 5 mg daily caused only a slight polyuria. The time of this experiment is too short to evaluate the effect on arterial pressure.

In a second series of experiments, the activity of compound S was tested in adrenalectomized rats and compared with that of DC. Briefly stated, the activity of compound S was estimated at about one-thirteenth that of DC in these tests, in which not only life maintenance, but also prevention of azotemia and of hypochloremia were compared at different dose levels.

We have concluded that it is unlikely that compound S as such is of major importance in the field of hypertension, and we are led to suggest that it is unlikely that it is a major factor in the maintenance of normal electrolyte metabolism. However, such activity as compound S showed was qualitatively similar to that of DC; hence, in view of the dosages employed, our data in no way negate the observations of Skelton and Selye reported above.

Clinical

You may recall that some of the hypertensive patients studied by Browne and Venning showed abnormally high titers of urinary glycogenic steroids and also that Dobriner and his colleagues found abnormalities of the urinary 17-ketosteroids in some, but not all, of the hypertensive patients they studied. Both sets of data suggest some hypertensive patients have an associated disorder of adrenal cortical function. Neither of these studies characterized the type of hypertensive disease observed or the effect of complications. Since also a few measurements of urinary corticoids including our own in hypertensives showed no abnormality, we were led to attribute such abnormalities as occurred to the effects of stress imposed by some complication of the disease, such as congestive failure.

The method we use (Corcoran and Page, 1948) is based on measurement of formaldehydogenic steroids. It is essentially that used by Dr. H. L. Mason. We have recently extended and re-examined our observations in normal subjects. These are summarized in Table 2 and show a significant difference between males and females, with higher values in males. For some unexplained reason, our normal values are about 0.2 mg higher per 24 hours than Dr. Mason's means.

TABLE 2

Urinary Corticoids in Hypertension

	No.	Mean	Range
Normal			
Male	12	1.1	(0.6—1.8)
Female	16	0.6	(0.2—1.4)
Patients			
Male	15	1.1	(0.2—1.9)
Female	15	1.4	(0.57—3.5)
Essential	10	1.27	(0.2—3.5)
Malignant	20	1.29	(0.33—2.8)

Of the 15 males, 3 had levels $>$ 1.8
Of the 15 females, 6 had levels $>$ 1.49
Of the 10 essentials, 3 had levels $>$ 1.8
Of the 20 malignants, 3 had levels $>$ 1.8

Comparing now our hypertensive patients with the normals, there is no difference in means between normal and hypertensive males. Hypertensive females do show a higher mean level of corticoid output, and the range is also beyond the normal in some hypertensive males. Thus we have to conclude that some hypertensives do show high urinary corticoids. But is this secondary and due to stress, to congestive failure, to some other complication, or to the rapid progress of vascular damage characteristic of malignant hypertension? We deliberately avoided doing determinations in the presence of recognizable stress other than that inherent in the disease. The point can be examined by comparing outputs in essential hypertension, which is not rapidly progressive, with those in malignant hypertension. If stress of disease were to account for the observed high outputs, these should predominate among the malignants. But they do not. Actually, the proportion of high outputs is greater in the essentials. Consequently, we conclude that the high urinary corticoid outputs of some hypertensives are not attributable to stress and that they may have some more fundamental association with the disease process.

The nature of this association is under study with particular attention to sodium retention and to the presence of sodium-retaining steroids and to possible excretion products thereof, such as pregnanediol. No data are yet available to suggest a correlation between any of these phenomena and the excretion of urinary corticoids.*

Still another fact which, as our understanding increases, may associate adrenal cortical function with essential hypertension, is that some patients respond to diets with a sodium content of less than 500 mg daily by a prolonged decrease in arterial pressure. The proportion of such responses is in our experience less than a third of patients rigidly observed. Such responses are much less common in out-patient work, where, of course, the diet cannot usually be adequately controlled. But the fact that arterial pressure and sodium intake show some association is reminiscent of DC hypertension in rats. Other observations which might link the adrenal to hypertension are the tendency of blood pressure to rise in some patients on very high sodium diets (Perera and Blood, 1947) and the abnormal tendency of some hypertensives to retain salt and water when suddenly deprived of dietary sodium (Perera and Blood, 1946).

The difficulties of group serial analyses prevent us from presenting still another fact which bears on this point as definitive or settled. The data at hand (Table 3) show that some patients who have abnormally high urinary corticoids in control periods on normal diets or diets of high sodium content respond to low sodium diets by concurrent decreases of urinary sodium and corticoid and of arterial pressure. Contrariwise, a few hypertensives whose corticoid excretions were normal in control periods show no change in this function, nor in arterial pressure, during sodium restriction. Indeed, one of them (Table 3) shows occasional increases of corticoid output which may be responses to the stress of severe sodium deprivation. Further experience is necessary before it can be concluded that the urinary corticoid level is a means of selecting patients for low sodium dietotherapy.

Recently, three patients with severe essential hypertension were observed during treatment with large doses of a concentrate of adrenal cortical extract dissolved in propylene glycol and generously furnished by Dr. H. Hailman of the Upjohn Co. During treatment, there was no consistent change in the urinary corticoids, which, however, were transiently decreased in two when treatment was stopped. The treatment period was associated with sodium retention and its interruption with sodium loss. Blood pressure and renal hemodynamics were unaltered in two patients during treatment. A decrease in blood pressure occurred in one patient but, with closer study, it seems to have been attributable to a complication of his disease.

Recently, two similar patients have been treated for 10 days with 100-mg daily dosages of cortisone and another two with 100-mg dosages of ACTH, the latter generously furnished by Dr. John R. Mote of the Armour Laboratories. One of the two patients treated with cortisone showed a significant decrease of arterial pressure during the first week of treatment. However, and it illustrates the pitfalls of short-term studies in small groups of the chronically sick, this decrease may have been due to a concurrent severe sodium restriction since it could not be reproduced in two subsequent test periods. No

* Added data have since been published, A. C. Corcoran, I. H. Page, H. P. Dustan, *J. Lab. Clin. Med.*, **36**, 297 (1950).

TABLE 3

A

C. L. Ess. Ht. Period	B.P. mm Hg	Sodium g	Corticoid mg
1	168/108	4.7	3.8
2	178/116	4.5	3.0
3	178/118	4.9	2.7
4	182/117	5.1	3.1
5	131/100	3.3	0.11
6	121/89	1.6	0.11
7	122/86	0.9	0.10
8	117/83	0.97	0.15

B

H.K.			
1	232/132	1.1	3.3
2	209/122	1.7	3.7
3	225/124	0.9	1.3
4	202/121	3.1	2.0
5	193/110	0.2	0.4
6	195/111	0.1	0.9
7	182/103	0.07	0.4
8	194/113	0.3	1.45
9	180/109	0.07	1.1

C

O. Ch. Ess. Ht.			
1	199/120	1.7	0.76
2	212/121	1.4	0.98
3	220/123	....	0.36
4	225/129	0.6	1.1
5	219/133	0.45	1.75
6	213/128	0.18	1.58
7	199/120	0.26	3.0
8	196/118	0.27	0.76
9	197/112	0.20	0.71
10	196/118	0.10	2.0
11	204/115	0.05	1.76

Note: A. Mean of successive weekly arterial pressures, urinary sodium (grams per 24 hours) and corticoids (milligrams per 24 hours) in responsive hypertensive patient C.L. B. In a less responsive patient H.K. C. In irresponsive patient O. Ch.

clinically significant change in arterial pressure occurred during treatment with ACTH. Renal hemodynamics and tubular excretory capacity for ρ-aminohippurate (Tm_{PAH}) were not altered by either cortisone or ACTH. Tubular reabsorptive capacity for glucose (Tm_G) was decreased in the two patients given ACTH and in one of the two on cortisone. Other analyses, including determinations of blood gluthathione by Dr. A. Lazarow, have still to be completed and correlated.

These studies with adrenal extracts, with cortisone and ACTH, were undertaken to investigate the possibility that the therapeutic effect (Page and Taylor, 1949) of pyrogens of bacterial origin in malignant hypertension might depend on altered adrenal cortical function. Preliminary observations (Corcoran and Page, 1948) had shown that the early phases of pyrogen treatment in patients who showed depressor responses to treatment were associated with increased excretion of urinary corticoid. In contrast, when pyrogen treatment had been prolonged for months and when the patients were resistant to its effects, corticoid output was not increased. Further study of this problem, as of the effects of adrenal-active materials, has been undertaken by Dr. Dustan. No definitive data are available from her study. A survey of what has been found with adrenal extracts, cortisone, and ACTH does not suggest that adrenal cortical activation is a basic mechanism of the therapeutic response to pyrogens. But the period of observation with these materials was brief, so that further, more prolonged studies of a similar nature are planned.

Summary

A review of the clinical data, like that of the experimental, suggests, but does not prove, that the adrenal cortex is genetically concerned in essential hypertension in human beings. Relevant to the problem are the facts that some hypertensives respond to low sodium diets, that some of these show abnormally high urinary formaldehydogenic corticoids in control periods, and that some show abnormal tendencies to sodium retention.

Brief periods of treatment with adrenal extract, with cortisone and ACTH have inconclusive effects on arterial pressure, although Tm_G is depressed by the two latter agents.

The therapeutic response to pyrogens of bacterial origin is associated with adrenal cortical activation, but whether concurrently or genetically remains to be established.

REFERENCES

Corcoran, A. C., and Page, I. H., *J. Lab. Clin. Med.*, **33**, 1326 (1948).

Corcoran, A. C., Taylor, R. D., and Page, I. H., *Ann. Internal Med.*, **28**, 560 (1948).

Lewis, L. A., and Page, I. H., *Federation Proc.*, **6**, 152 (1948).

Masson, G., Corcoran, A. C., and Page, I. H., *J. Lab. Clin. Med.*, **34**, 1416 (1949).

Masson, G., Hazard, B., Corcoran, A. C., and Page, I. H., *Arch. Path.* (submitted for publication), 1950.

Masson, G., Corcoran, A. C., and Page, I. H., *Endocrinology*, (submitted for publication), 1950.

Page, I. H., and Taylor, R. D., *Modern Concepts of Cardiovascular Disease*, **18**, 51 (1949).

Perera, G. A., and Blood, D. W., *J. Clin. Invest.*, **26**, 1109 (1947).

Perera, G. A., and Blood, D. W., *Am. J. Med.*, **1**, 602 (1946).

Selye, H., *J. Clin. Endocrinol.*, **6**, 117 (1946); *A Textbook of Endocrinology.* Montreal, 1947.

THE ADRENAL CORTEX IN DIABETES MELLITUS*

EDGAR S. GORDON
University of Wisconson, Madison

Because of the important role played by the adrenal cortex in the regulation of carbohydrate metabolism, it is quite natural that the participation of this gland in the disease complex of diabetes mellitus should be subjected to considerable investigation. The adrenal cortex is known to exert a significant influence upon the conversion of protein into carbohydrate, as demonstrated in 1940 by Long, Katzin, and Fry, and through that mechanism to provide a controlling action over the availability and supply of carbohydrate as a component of the metabolic mixture that is available for oxidation, over the quantity of glycogen present in the liver and over the level of glucose in the blood and tissues. The cortex also exerts a control over the oxidation of carbohydrate in the peripheral tissues, and although this effect is poorly understood, it is probably mediated through an increase in the resistance of those tissues to the action of insulin. If such a mechanism eventually proves to be correct, it is likely that it operates by means of an influence on phosphorylating reactions. From these facts it is quite apparent that fluctuating intensities of adrenal cortical function should be attended by parallel variations in the severity of the clinical manifestations of the diabetic state. Furthermore, increased activity of the adrenal cortex as regards its carbohydrate-regulating hormones might logically be expected to produce clinical diabetes, at least in a temporary and possibly even in a permanent form. Such a condition has been termed "adrenal diabetes," of which the metabolic pattern of Cushing's disease is probably the best type example. There is reason to believe, however, that adrenal function may play some part in the development and course of many cases of diabetes mellitus of the usual type.

In 1944 Venning, Hoffman, and Browne reported the quantitative measurement of adrenal cortical hormones in human urine, and since that time numerous investigators have demonstrated an increased excretion of these steroid compounds in response to infections, trauma, and stress in human subjects (Josiah Macy Jr. Foundation, 1945; Heard *et al.*, 1946; Venning and Kazmin, 1946). Superficially these findings would appear to provide a clear indication of increased production of these materials by the adrenal cortex, but despite some confirmatory evidence for this point of view, the possibilities of decreased tissue utilization of steroid hormones and a shift in quantitative synthesis of one or several steroid compounds over others must still be considered in seeking an explanation for this phenomenon. This altered pattern of steroid excretion is attended by the appearance of an aug-

* These investigations were made possible by means of funds supplied through a grant from the United States Public Health Service.

mented excretion of urinary nitrogen and by elevated blood sugar levels. Similar changes had been demonstrated even earlier in rabbits exposed to low oxygen tension. This response to nonspecific traumatic episodes in now a familiar pattern in animal physiology and has been suggested by Albright (1942-43) and others as the probable explanation for the upheaval in metabolic

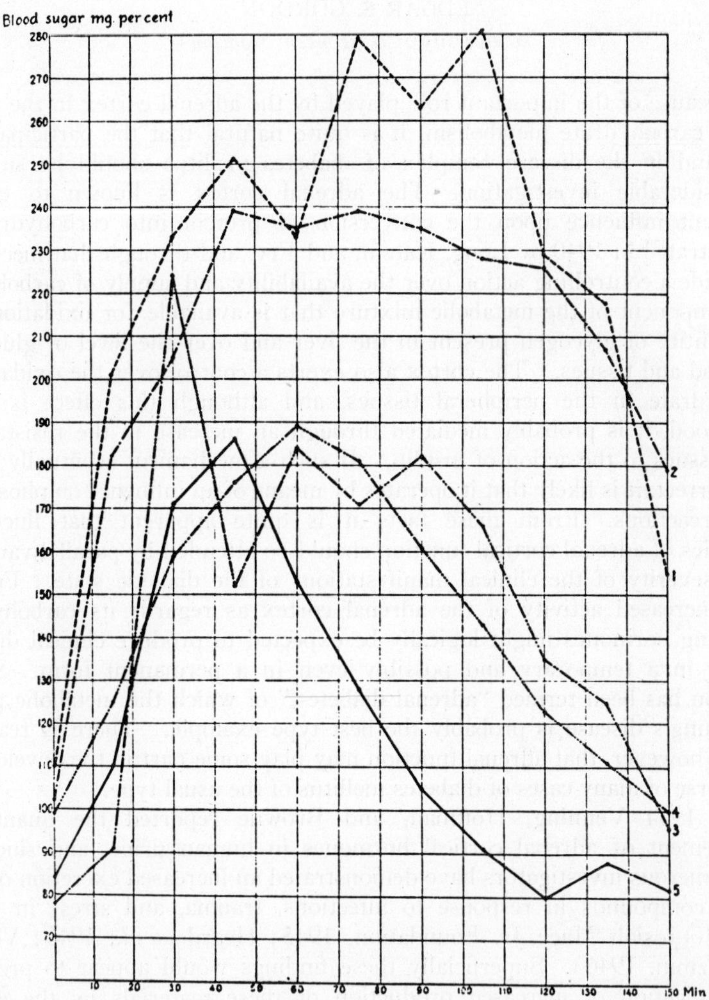

FIG. 1. Glucose tolerance curves in an otherwise healthy subject who suffered a severe contusion. After J. Thomsen, *Acta Med. Scand.*, **Suppl.** 1938.

balance which occurs regularly in previously well-controlled clinical cases of diabetes mellitus in response to infections, fractures, burns, surgical operations, etc. Thomsen in 1938 reported impairment of glucose tolerance as a nearly routine occurrence in fracture cases, in whom a temporary or permanent diabetic state developed in a significant number of instances, presumably as a result of this metabolic disturbance mediated through the endocrine system.

It seemed for a considerable time that the increased glyconeogenesis from protein clearly indicated by the chemical data of hyperglycemia, glycosuria, and negative nitrogen balance was actually caused by the metabolic effect of increased glycogenic steroid production by the adrenal cortex. Ingle *et al.* (1948) however, have shown that although these changes are coexistent, the negative nitrogen balance caused by an increased catabolism of protein, a diminished anabolism of protein or both, can and does occur even in the

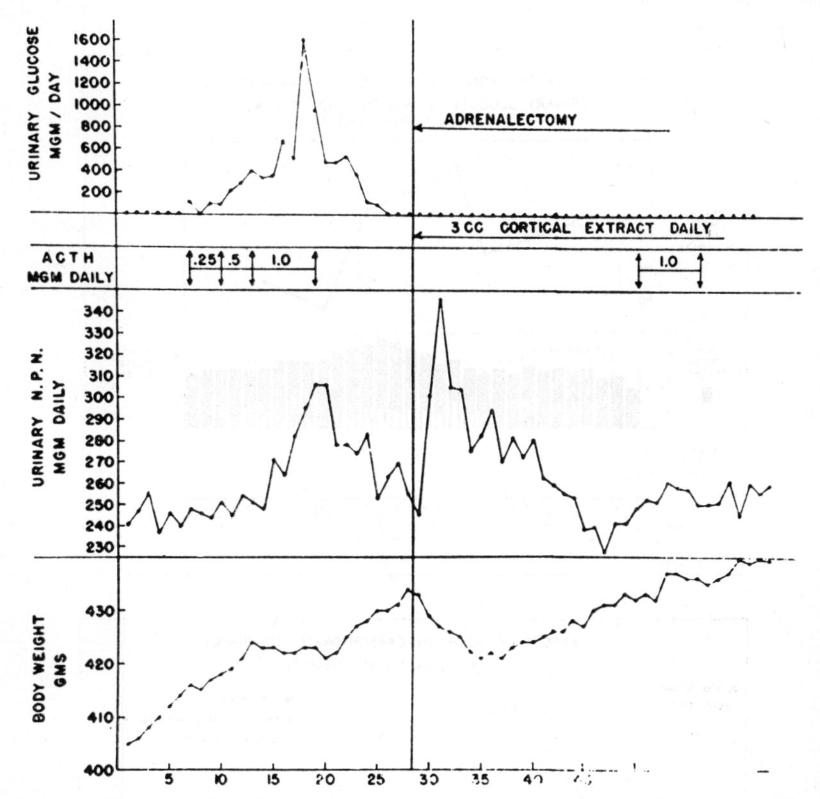

FIG. 2. Effect of adrenalectomy on urinary glucose and nitrogen in rats treated with ACTH. From D. J. Ingle *et al.*, *Endocrinology*, **43**, 202 (1948).

absence of the adrenal glands but not in the absence of the adrenal cortical hormones. Although no investigator to my knowledge has yet found a satisfactory explanation for these facts, there is some reason to believe that the presence of the thyroid hormone may also be necessary for the increase in nitrogen loss in the urine.

Inasmuch as this pattern of behavior is a regular occurrence in otherwise healthy animals in response to damage, a disturbance in protein metabolism and a rise in blood sugar level may be an equally regular occurrence in response to even such trivial insults as colds, minor operations, or even emotional disturbances. Whether or not it is present in measurable degree may depend upon the functional adequacy of the pancreatic islet cell response to the increased load imposed upon that gland. In the presence of a low reserve, pancreas "decompensation" may occur with the end result of either temporary or permanent clinical diabetes. In other words, the adrenal cortex may function, in the sense outlined, as a diabetogenic factor not only of theoretical, but of considerable practical importance as well. In individuals who have inherited a low reserve pancreas, the prophylactic use of insulin to cover periods of physiological strain attending trauma and infections may be a wise measure of prevention.

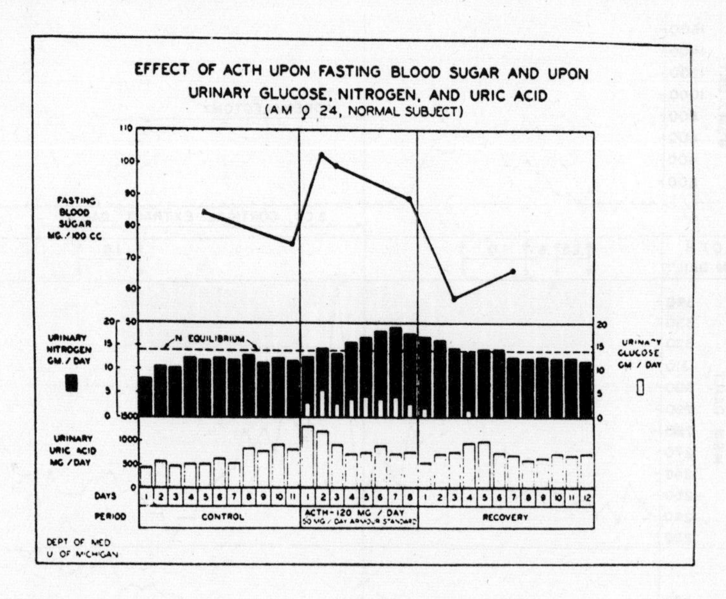

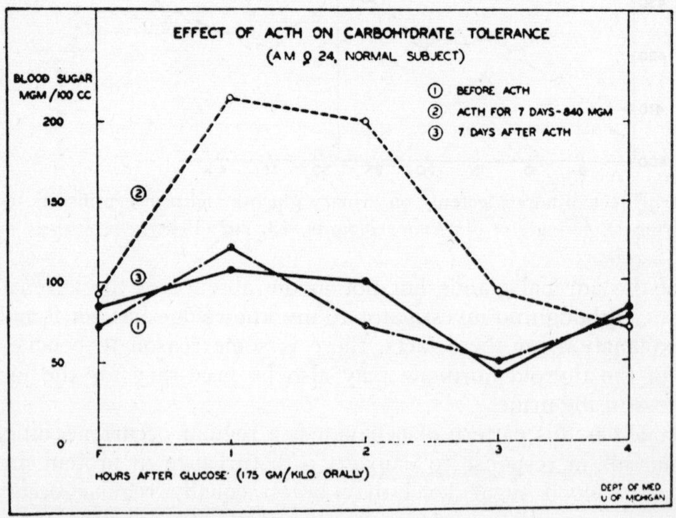

FIG. 3. Effect of ACTH in normal human subjects. From J. W. Conn, Proceedings American Diabetic Association, 1949.

Adrenal effects upon metabolism have also been studied successfully by means of the administration of both cortisone and ACTH. Conn (1949) has shown the hyperglycemic effect of ACTH in normal human subjects, and from his data it is clear that the maximum effect upon blood sugar levels and glucose tolerance curves is seen in those individuals who have shown also

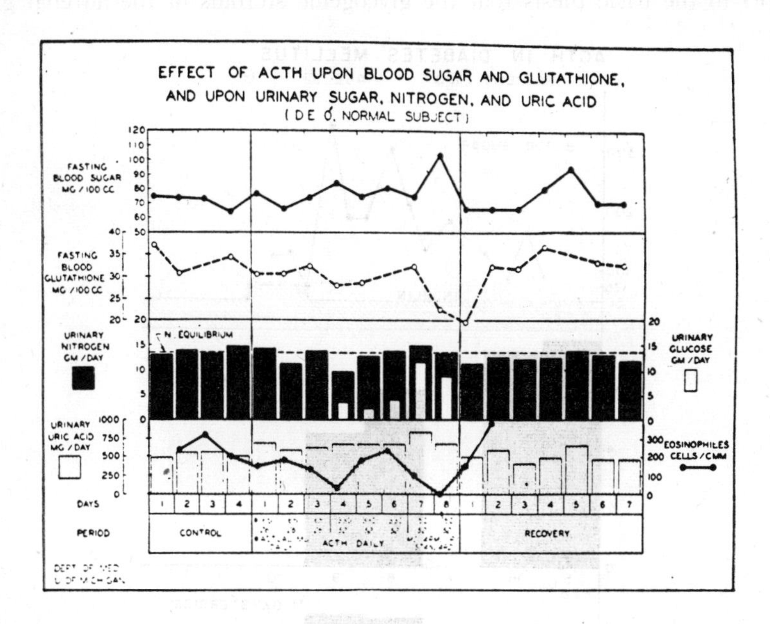

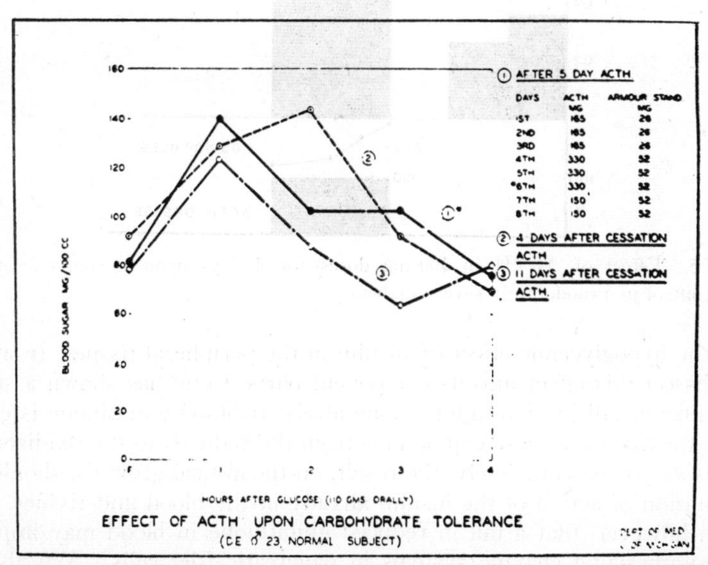

FIG. 4. Effect of ACTH in normal human subjects. From J. W. Conn, Proceedings American Diabetic Association, 1949.

a negative nitrogen balance. We have obtained almost identical results and have demonstrated, in addition, that ACTH makes clinical diabetes, already existent, more severe, in correction of which larger doses of insulin are necessary. It is of the greatest interest, however, that the diabetes so produced or so aggravated is often extremely insulin resistant — a fact which lends strong support to the basic thesis that the glycogenic steroids of the adrenal gland

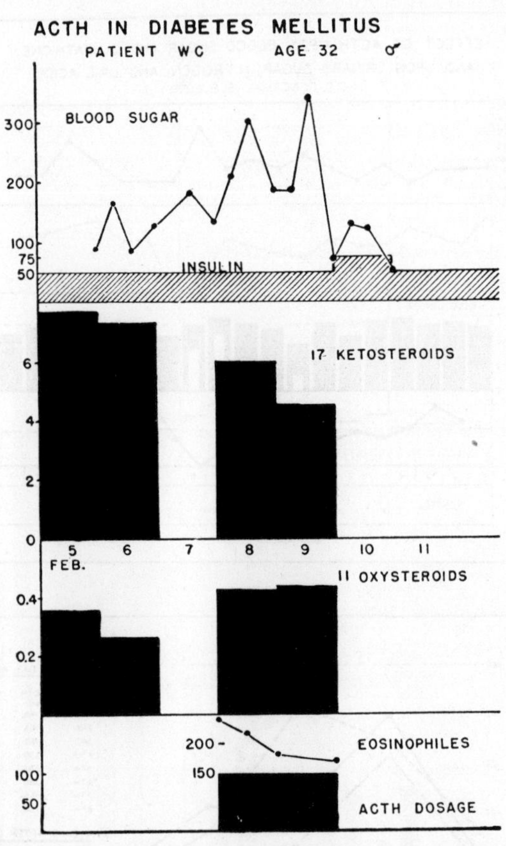

FIG. 5. Effect of ACTH in 100-mg doses for 2 days upon steroid excretion and diabetic control in a moderately severe diabetic.

reduce the hypoglycemic effect of insulin in the peripheral tissues. In attempting to dissect this effect into its component parts. Conn has shown a statistically significant fall in blood glutathione levels. If blood glutathione is changed through the action of these compounds from the reduced to the oxidized form, or if it is destroyed completely, the result, on theoretical grounds, should be an augmentation of action of the insulin already in the blood and tissues. Conn believes, however, that a fall in reduced glutathione in blood may impair the insulin-synthesizing enzyme systems in pancreatic islet cells. We also have been studying glutathione in whole blood using a different method (the electrometric titration method of Benisch) and the changes recorded are inconsistent and not of significant magnitude. This failure of agreement may be entirely a matter of methodology, but it is clear that a great deal more work is necessary, preferably by means of the glyoxylase method, the only one known to be specific for glutathione.

Attracted by the extreme instability of the diabetes in some human subjects, especially in those with advanced vascular degenerative changes being maintained on constant diet and insulin dosage, it seemed that several possible explanations might be formulated:

1. Sharp changes were occurring over periods of minutes or hours in insulin sensitivity. If this were true the cause of such changes most probably involved the activity of either the adrenal cortex or the pituitary gland.
2. Sharp fluctuations in available carbohydrate were occurring, probably as a result of erratic rates of glyconeogenesis from protein. Such conditions might also involve the adrenal cortex.
3. Some unknown hyperglycemic factor or factors (excluding epinephrine) were present in variable concentration, as, for example, a possible "alpha cell hormone."
4. A combination of these factors.

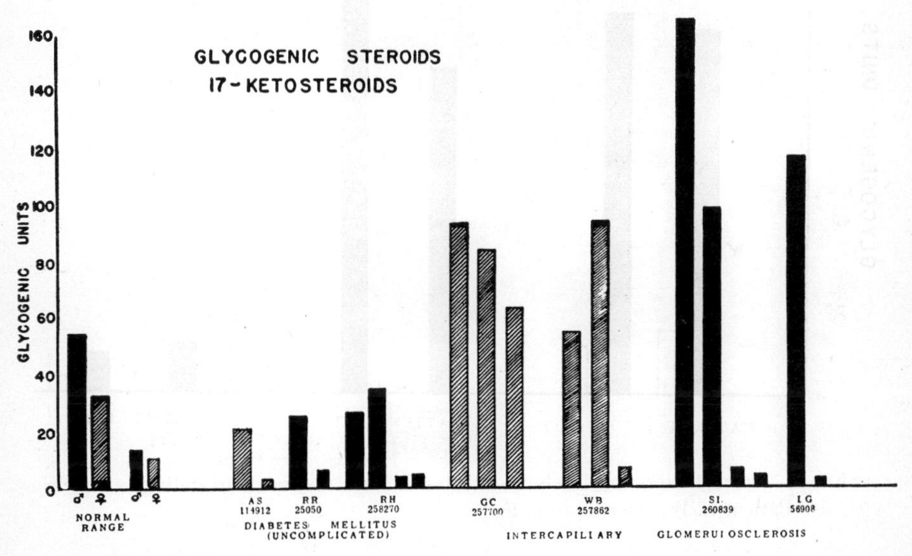

FIG. 6. Steroid excretion levels in diabetes mellitus and in intercapillary glomerulosclerosis. Patient L. W. referred to in text.

Obviously it seemed important to study quantitative adrenal cortical function in such cases. Accordingly, in diabetic patients of various clinical types rather comprehensive metabolic studies have been carried out with special attention paid to those individuals with advanced degenerative vascular disease and especially to cases diagnosed clinically as probable Kimmelstiel-Wilson intercapillary glomerulosclerosis. High levels of 11-oxysteroid excretion in diabetic coma have already been reported by Venning and others.

Normal levels of excretion have been found by several investigators in ordinary diabetics under reasonably good control. We have confirmed these findings, but in addition, we have been surprised at the unusually high levels of excretion of glycogenic steroids and the very low levels of excretion of 17-ketosteroids in cases of intercapillary glomerulosclerosis. Indeed, many of these patients are pouring out glycogenic steroids at rates in excess of those usually seen in Cushing's disease. The diagnosis in several but not in all the cases studied have been confirmed at autopsy. That this is actually a post-mortem rather than a clinical diagnosis is illustrated by one of our cases as shown in Fig. 7. This man, a diabetic of only moderate severity but of long

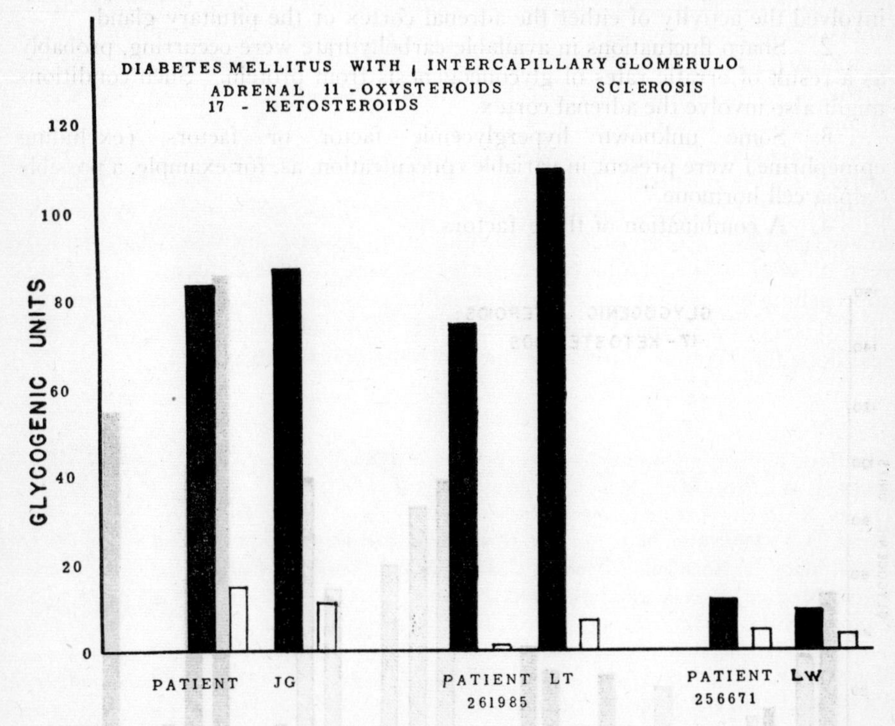

FIG. 7. Steroid excretion levels in diabetes mellitus and in intercapillary glomerulosclerosis. Patient L. W. referred to in text.

duration, had massive edema, hypertension, severe retinopathy, renal failure, and all the clinical findings characteristic of intercapillary glomerulosclerosis, but our excretion figures for 11-oxysteroids were within the normal range. At autopsy he was found to have nephrosclerosis with severe renal damage compatible with the clinical picture, but no sign of the characteristic lesions of intercapillary glomerulosclerosis.

The extent and duration of this study does not permit any estimate at the present time of the invariability of this pattern of steroid excretion, but it appears to be quite constant and unaltered by therapy. In attempting to attach some significance to these findings it is important to state that we have no evidence whether these endocrine changes are of primary etiological importance in the development of the vascular lesions or whether they are secondary in response to the vascular disease. That it is not due to or associated with the hypertension *per se* is indicated by the normal levels we and others have obtained in essential hypertension, including some cases of the malignant variety. Many of the individuals in this study were not in a terminal phase of their illness, so that the findings can hardly be regarded as a terminal phenomenon. We have tested the effect of testosterone, with a resulting rise in 17-ketosteroid and, in some instances, a fall in 11-oxysteroid excretion. Nitrogen balance has been strongly positive in all these subjects presumably because they had all been under uniformly poor control before admission to the hospital for study. Under such conditions any improvement in management results in clinical improvement which leads to a storage of nitrogen. For this reason we do not consider the nitrogen balance data to be of any great significance.

In summary, there is abundant evidence to indicate that the adrenal cortex is of fundamental importance in the metabolic complex of diabetes mellitus. That disease which seemed for a brief time to be so near to being a solved problem at the time that insulin first became available, has yielded very slowly to the combined efforts of scores of investigators. Anything approaching a complete understanding still seems far off, but increasing attention to the complete biochemical integration of the endocrine system provides real hope for solution of many of the problems which still remain.

REFERENCES

ALBRIGHT, F., Cushing syndrome, *Harvey Lectures 1942-43.*

CONN, J. W., *Proceedings American Diabetic Association,* 1949.

HEARD, R. D. H., SOBEL, H., and VENNING, E. H., *J. Biol. Chem.,* **165,** 699 (1946).

INGLE, D. J., PRESTRUD, M. C., and LI, C. H., *Endocrinology,* **43,** 202 (1948).

Josiah Macy Jr. Foundation Reports, *Conference on Metabolic Aspects of Convalescence,* 10th Meeting, New York, June 15-16, 1945.

LONG, C. N. H., KATZIN, B., and FRY, E. G., *Endocrinology,* **26,** 309 (1940).

THOMSEN, J., *Acta Med. Scand.,* **Suppl.,** 1938.

VENNING, E. H., HOFFMAN, M. M., and BROWNE, J. S. L., *Endocrinology,* **35,** 49 (1944).

VENNING, E. H., and KAZMIN, V. E., *Endocrinology,* **39,** 131 (1946).

PITUITARY-ADRENOCORTICAL FUNCTION IN PATIENTS WITH SEVERE PERSONALITY DISORDERS

HUDSON HOAGLAND

The Worcester Foundation for Experimental Biology, the Worcester State Hospital, and the Department of Physiology, Tufts College Medical School, Shrewsbury, Massachusetts

This paper is a brief summary of the work of our group in the study of adrenal cortical function in "mental" patients. It is a survey of many published papers and of work now under way. In form it is similar to our paper given at the Armour Conference on ACTH in October, 1949.

Engaged in these studies in the past six years, in addition to the author have been Gregory Pincus, Harry Freeman, Fred Elmadjian, Louise Romanoff, William Malamud, Sidney Sands, Enoch Callaway, Justin Hope, Victor Schenker, Eliot Rodnick, Austin Berkeley, James Carlo, and some twenty technicians without whose help the work would not have been possible.

Our investigations have been aided by the Office of Naval Research, the U. S. Public Health Service, the Williams-Waterman Fund of the Research Corporation, the Scottish Rite Committee for Research on Dementia Praecox, the F. C. and B. Foundation, Armour and Company, and the G. D. Searle Company.

Chronic schizophrenic patients, not subjected to special stresses, give evidence of adrenal cortical secretion differing little from normal controls. Thus we have found that the 24-hour excretion of 17-ketosteroids, potassium, and uric acid in 34 patients and 36 controls fasted overnight is not different, although a significantly smaller excretion of neutral reducing lipids was observed in the patient group. The patients also excrete significantly more sodium and more urine, but water intake was not adequately controlled and these last results are hard to evaluate (Pincus, Hoagland, Freeman, Elmadjian, and Romanoff, 1949; Pincus, Hoagland, Freeman, and Elmadjian, 1949).

We have studied adrenal cortical function in approximately 100 psychotic patients and 200 control subjects in a variety of stressful tests. These tests have been exposures to heat and high humidity (Pincus and Elmadjian, 1946), exposure to cold (Pincus, 1947), the fatiguing operation of an airplane-type pursuit meter at sea level and at reduced oxygen tensions (Pincus and Hoagland, 1943; Hoagland, Elmadjian, and Pincus, 1946), the stress of ingesting large doses of sugar (Elmadjian, Freeman, and Pincus, 1946; Freeman and Elmadjian, 1946, 1947), psychological tests such as interviews, and a specially designed frustration test (Pincus and Hoagland, 1943; Hoagland, 1947; Pincus, Hoagland, Freeman, Elmadjian, and Romanoff, 1949). Urine and blood samples were taken before each test, which lasts an hour, 15 minutes after the test, and again at two hours and 15 minutes post-test. The urinary samples were analyzed for 17-ketosteroids, neutral reducing lipids (cortins), sodium, potassium, uric acid and inorganic phosphate. Lymphocyte counts were made on the blood samples and also eosinophil counts in many of the tests.

Not all the subjects have been through all our tests and studied by all our indices, but our findings have been evaluated statistically and may be summarized by saying that the patient group (schizophrenics hospitalized on an average of 2½ years) failed markedly to enhance adrenal cortical output with

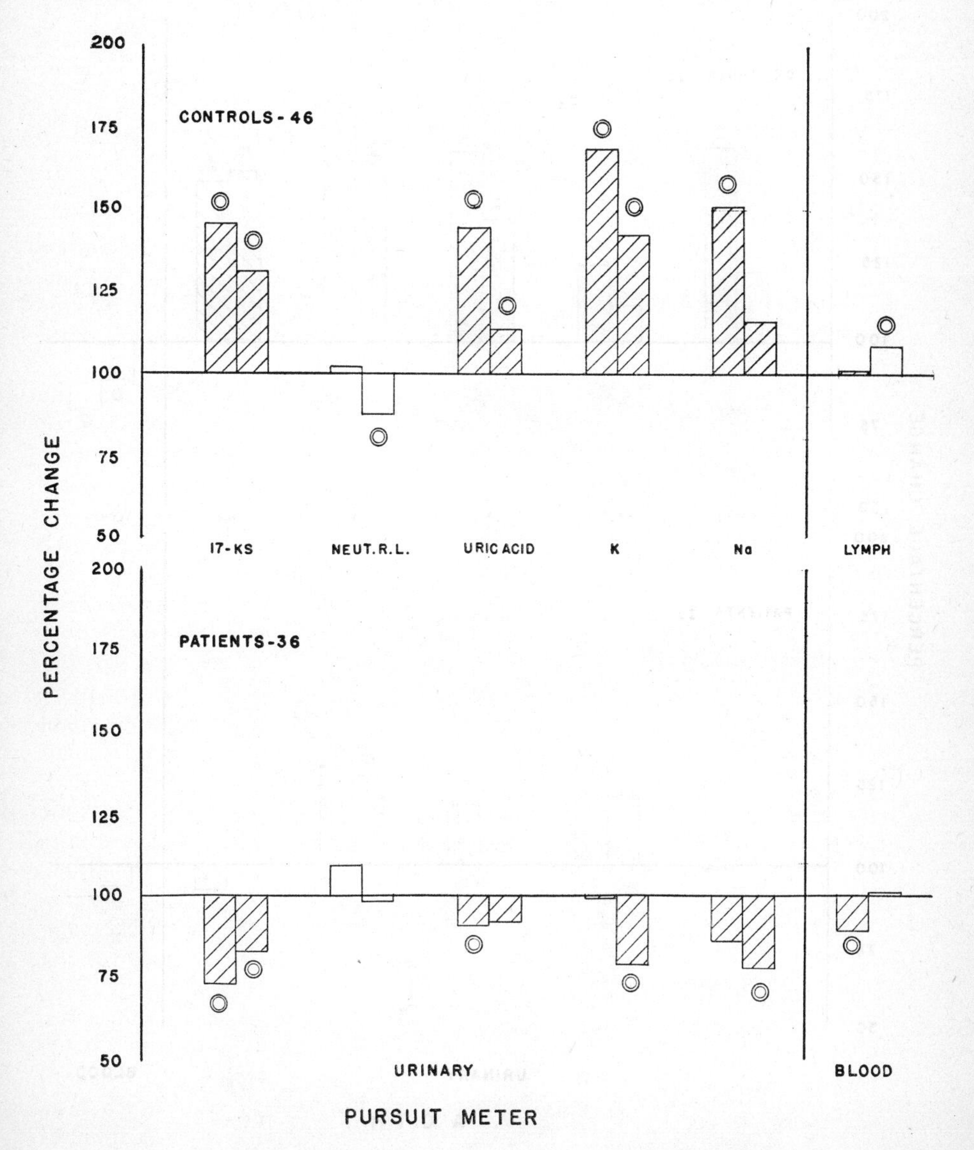

FIG. 1. Comparisons of six adrenal cortical stress response measures in response to pursuit meter tests given to 46 fasting controls and 36 fasting schizophrenic patients. For discussion, see text.

stress as compared to the control group. This is true especially for purely physiological stresses as well as for psychological stresses. Figure 1 shows differences in responses to an hour of operation of a pursuit meter by 46 controls and 36 patients. The first rectangle of each pair represents per cent changes from pretest levels at 15 minutes post-test, and the second rectangle

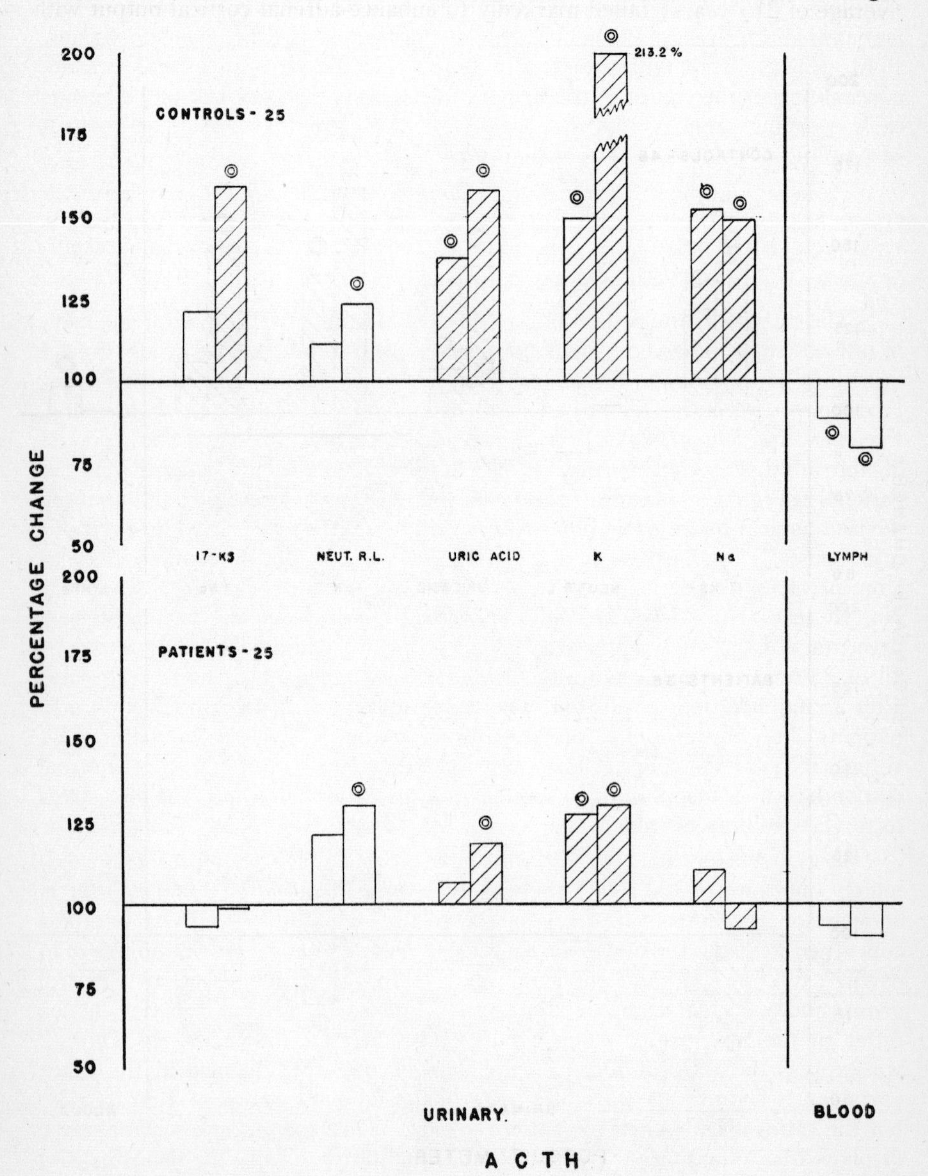

Fig. 2. Comparisons of six adrenal cortical response measures following injection of 25 mg of ACTH to 25 fasting controls and 25 fasting schizophrenic patients. For discussion, see text.

shows changes at 2¼ hours post-test. Circles above the rectangles represent changes from resting levels at better than the 5% level of confidence and cross hatchings indicate differences between patients and controls at this same level of confidence. Control urine and blood samples were collected over the first 2 hours after rising, and the post-test samples were late morning samples. We have demonstrated a diurnal rhythm of adrenal secretion which is maximal after rising and which falls during the day (Pincus and Hoagland, 1943; Pincus, Romanoff, and Carlo, 1948). If the stress does not enhance adrenal output the values of the urinary variables should fall below the resting early morning samples, and this they do in the patient group. The normals responding to the stress show enhancement of the urinary indices.

Plots similar to those of Fig. 1 have been made for the stress of our "targetball" frustration test and for the ingestion of sugar (Pincus and Hoagland, 1950; Freeman and Elmadjian, 1950). Data summarizing results of our various stress tests will now be discussed.

The target organs responding to adrenal steroids are normally responsive in our patients, as has been demonstrated in control and patient groups by the injection of standard doses of Upjohn's lipo-adrenal extract (Pincus, Hoagland, Freeman, Elmadjian, and Romanoff, 1949; Pincus, Hoagland, Freeman, and Elmadjian, 1949; Pincus and Hoagland, 1950). The response of the adrenal cortex to ACTH is, however, abnormally sluggish in the patients (Fig. 2). A total response index (abbreviated TRI) has been devised as a measure of adrenal responsivity to our stresses and to hormone injections. This consists of a mean of the sum of the percentage changes from prestress or pre-injection levels of the two poststress samples of urinary Na, K, uric acid, 17-ketosteroids, and cortins plus twice the two poststress percentage drops in lymphocytes added as positive numbers. For example, all of 25 normal control subjects showed a TRI index of 20 units or greater after 25 mg injections of ACTH, whereas only 28% of 25 of our schizophrenic patients showed a response of 20 units or greater. Table 1 compares TRI values in some of our tests. Approximately this same ratio of normal responders to schizophrenic responders is also found for our various stress tests (last column of table).

In the schizophrenic group who gave *normal* TRI responses (approximately one-third of the total) there is evidence of *qualitative abnormalities* in comparison to controls in secretion of relative amounts of adrenal steroids concerned with specific metabolic functions. We found no stress % difference in cortin excretion (neutral reducing lipides) in the patient from the control group and we have compared our other indices of adrenal function which differ in the two groups using cortin excretion as a standard. In Table 2 we express the ratio of mean cortin changes to mean changes in the other variables. The urinary variables in normal men show ratios of 1.30 to 1.38, but the ratios are lower in the patient group. Thus the chronic schizophrenic displays both quantitative (Table 1) and qualitative (Table 2) abnormalities of adrenal stress responses which appear to lie specifically at the level of the ability of his adrenal cortex to respond to ACTH by release of normal amounts and normal ratios of steroid hormones.

TABLE 1

The Total Response Index of Various Subjects in the Stress Tests

Test	Subjects	Mean TRI	% of Subjects with Score of 20 or Greater	Ratio of Normal Responders to Schizophrenic Responders
ACTH	Normal (25)	46.2	100.0	
ACTH	Schizophrenic (25)	12.0	28.0	3.6
Pursuit	Normal (46)	22.1	47.8	
Pursuit	Schizophrenic (36)	—2.7	15.8	3.0
Targetball	Normal (36)	11.5	32.5	
Targetball	Schizophrenic (20)	—6.7	10.0	3.3
Glucose tolerance	Normal (47)	22.1	48.6	
Glucose tolerance	Schizophrenic (38)	5.7	15.0	3.2

The abnormal responses are not corrected by a vitamin and protein rich diet (Pincus, Schenker, Elmadjian, and Hoagland, 1949). They are not found in psychoneurotic patients who we have found to give normal responses to 25-mg injections of ACTH (Pincus, Hoagland, Freeman, and Elmadjian, 1949; Pincus and Hoagland, 1950).

The refractoriness of the adrenals to ACTH in the psychotic patients displaying abnormal responses is not absolute. Thus larger quantities of ACTH (75 to 100 mg) produce TRI responses more nearly approaching those of the controls given 25-mg doses (Pincus and Hoagland, 1950).

We have reported that normal persons show an increase of 50±14% in output of 17-ketosteroids in the first 2 hours after rising after a night's sleep (Pincus and Hoagland, 1943; Pincus, Romanoff, and Carlo, 1948). This reflects a special form of adrenal stress response associated with waking and

TABLE 2

The Ratio of Mean Percentage Change in Neutral Reducing Lipid Output to the Mean Percentage Change in Other Indices

The data are those of the fourteen schizophrenic men classified as positive responders (TRI>20) to the stress tests and a similar group of normal subjects.

Response Index	Ratio	
	Schizophrenic Men	Normal Men
17-ketosteroid output	1.12	1.35
Uric acid output	0.80	1.38
Potassium output	0.87	1.31
Sodium output	0.94	1.30
Lymphocyte number	0.61	0.82

starting the day's activities. In general, schizophrenic patients display, as a group, a somewhat lower diurnal rhythm of excretion than do normal controls (Pincus, 1947).

In ten female patients suffering from involutional depression we also found evidence of adrenal cortical unresponsivity (Pincus, Schenker, Elmadjian and Hoagland, 1949). These patients averaged an insignificant (3%) morning increase in 17-ketosteroid excretion over the night level before they were given a course of electroshock treatments. During the treatment course,

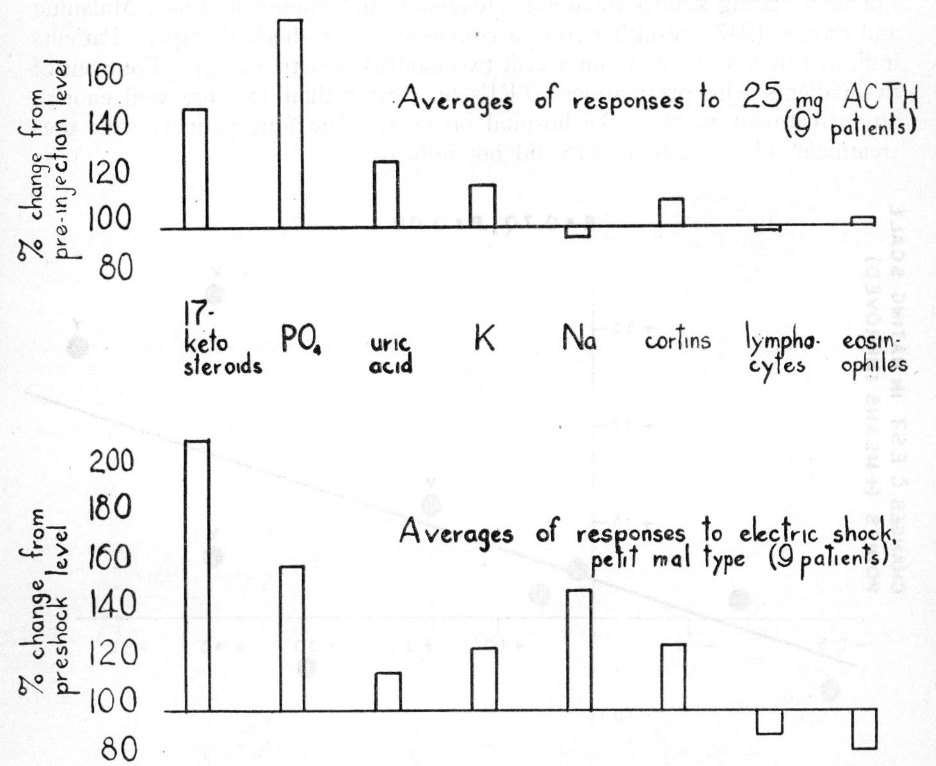

FIG. 3. Comparison of adrenal stress responses in the same patients when injected with 25 mg of ACTH and when given non-convulsive electroshock treatments. For discussion, see text.

the morning increase in output rose to an average 32% over the night level, and it was 25% over the night level a month after the treatments had ended. Eight of the ten women showed good social remission resulting from the treatments. Thus we see that adrenal unresponsivity to stress is found in psychotics other than schizophrenics and that the condition improves with therapy.

In a recent series of experiments on nine schizophrenic patients we have demonstrated that electric shock stimulates adrenal activity to a degree comparable to the injection of approximately 100 mg of ACTH (Hoagland, Callaway, Elmadjian, and Pincus, 1950). Figure 3 compares the response of these patients an hour and a half post-injection of 25 mg of ACTH with the response to electric shock across the head sufficient to produce unconsciousness, but not strong enough to bring about major convulsion.

We have also found (Hoagland, Callaway, Elmadjian, and Pincus, 1950) that favorable prognosis in electroshock therapy of schizophrenic patients is correlated with the degree of responsivity of the adrenal cortex to pretreatment doses of 25 mg of ACTH ($r = 0.7$, $P = 0.02$). Figure 4 is a plot of the TRI (hoizontally) against the total number of units of improvement on the Malamud rating scale (Malamud, Hoagland, and Kaufman, 1946; Malamud and Sands, 1947) resulting from a course of electroshock therapy. Patients indicated by a V were out on a visit two months post-treatment. Four out of five patients with pretreatment TRI's of greater than 15 were well enough after treatment to leave the hospital on visit. The four patients with pretreatment TRI values of <15 did not improve.

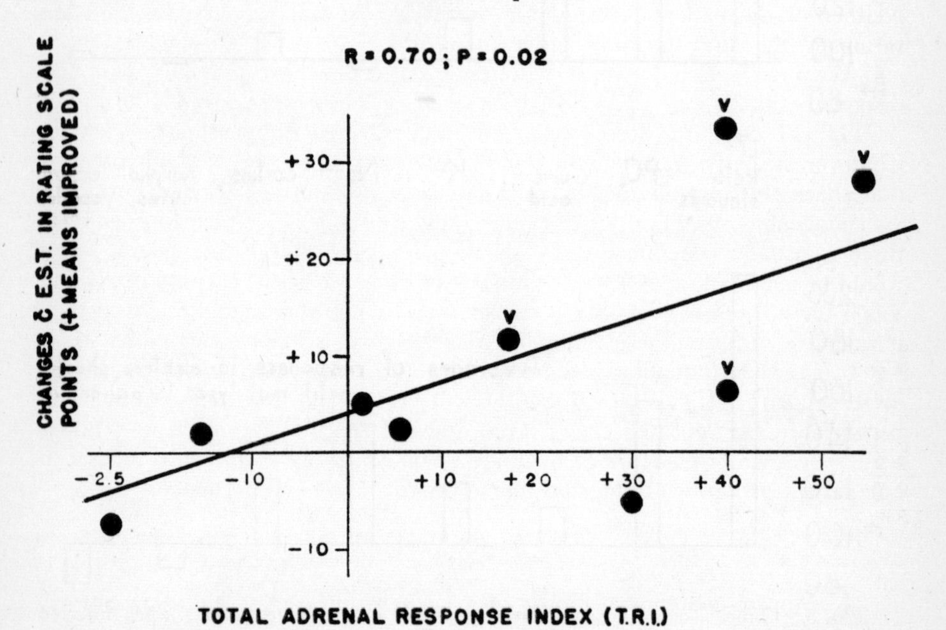

FIG. 4. This figure indicates that the better the response of the patient's adrenals to a standard dose of ACTH, the better is his prognosis when subsequently given electroshock therapy. For discussion, see text.

The findings suggest several hypotheses. It is possible that all shock therapies (electric, insulin, and metrazol) act as repeated violent stresses to discharge excessive amounts of endogenous ACTH which thereby activates the patients' sluggish adrenals inducing consequent beneficial metabolic changes. To test this hypothesis we have first selected five schizophrenic patients with relatively responsive adrenals, who, from our findings, would be expected to do well on shock. We have given these patients daily injection of ACTH of from 100 to 200 mg. Four patient controls receiving placebos were also studied in this series. All the patients were followed by ratings on the Malamud scale.

One of the patients had to be dropped from our study after one week owing to complaint from the family. Another of this group was discontinued after a week of injections because of cardiac complications, and only three of the five were carried through three full weeks of medication. The total psychiatric index registered no significant improvement in two of the patients, but one showed marked psychiatric improvement in the second and third weeks (100 mg per day). He became less seclusive, played games with other patients and engaged in ward work. He relapsed the week following treatment.

The other two patients treated for three weeks showed no net overall improvement though one became euphoric and the other, in the second and third weeks, had improvement in insight, and his affect went from flat to angry and explosive. His mood changed from optimistic to depressed. Neither of the patients treated for one week showed changes, nor did the ratings of the controls vary significantly.

We next selected two schizophrenic patients whose adrenals were unresponsive to 25 mg of ACTH (TRI $<$20). They were given 100 mg of ACTH per day for three weeks, but they showed no psychiatric changes whatever. Negative results with so few patients do not disprove the hypothesis that shock therapy may act by releasing large amounts of adrenal hormones via endogenous ACTH discharge since only about one of our schizophrenics in three is usually benefited by the shock therapy. A large therapeutic series would be necessary to test the hypothesis and because of the scarcity of ACTH this is not at present practicable. Even if the hypothesis should prove correct, it would be of little practical importance in treating schizophrenic patients.

It may be that qualitative abnormalities in adrenal output in schizophrenics in response to ACTH (see Table 2) will prevent this substance from being beneficial. The repeated use of large doses of specific steroids is now under way. Because of evidence of qualitative abnormalities of adrenal function in our patient group, we think that this approach may be more hopeful.

Summary

1. Approximately two-thirds of a large group of chronic schizophrenic patients show unresponsivity of the adrenal cortex to several psychological and physiological stresses in contrast to normal controls. Six urinary and blood indices were used as measures of the adrenal responses.

2. Test injections of adrenal cortical extract produce similar responses of the indices in control and patient groups indicating that target organs acted upon by the adrenal steroids respond normally.

3. Two-thirds of the schizophrenic patients are unresponsive, by our criteria, to 25-mg test injections of ACTH as compared to our controls indicating that the unresponsivity lies at the level of the target organ for ACTH, i.e., the adrenal cortex itself. Large doses of ACTH do, however, activate the patients' adrenals somewhat more effectively. The unresponsivity of the patients is not corrected by a diet rich in proteins and vitamins.

4. Psychoneurotic patients display normal responses to ACTH.

5. One-third of the schizophrenic patients give adrenal responses to our stresses and to ACTH within the normal range as measured by an overall adrenal response index which is the mean of the sum of the changes of the six urinary and blood variables. This group, however, shows qualitative abnormalities of response compared to normal persons in terms of the ratios of their various urinary indices.

6. Electroshock stimulates the pituitary-adrenal system in the schizophrenic patients, and those patients with the more responsive adrenals to test doses of ACTH improve more on subsequent electroshock therapy than do those with unresponsive adrenals.

7. ACTH in daily doses of from 100 to 200 mg administered for 3 to 4 weeks was not a therapeutic agent in five patients so far treated.

REFERENCES

ELMADJIAN, F., FREEMAN, H., and PINCUS, G., The adrenal cortex and the lymphocytopenia due to glucose administration. *Endocrinology,* **39,** 5 (1946).

FREEMAN, H., and ELMADJIAN, F., The relationship between blood sugar and lymphocyte levels in normal individuals. *J. Clin. Endocrinol.,* **6,** 668 (1946).

FREEMAN, H., and ELMADJIAN, F., The relationship between blood sugar and lymphocyte level in normal and psychotic subjects. *Psychosomat. Med.,* **9,** 226 (1947).

FREEMAN, H., and ELMADJIAN, F., *Am. J. Psychiat.,* **106,** 660 (1950).

HOAGLAND, H., ELMADJIAN, F., and PINCUS, G., Stressful psychomotor performance and adrenal cortical function as indicated by the lymphocyte response. *J. Clin. Endocrinol.,* **6,** 301 (1946).

HOAGLAND, H., MALAMUD, W., KAUFMAN, I. C., and PINCUS, G., Changes in the electroencephalogram and in the excretion of 17-ketosteroids accompanying electroshock therapy in agitated depressions. *Psychosomat. Med.,* **8,** 246 (1946).

HOAGLAND, H., The human adrenal cortex in relation to stressful activities. *J. Aviation Med.,* **18,** 450 (1947).

HOAGLAND, H., CALLAWAY, E., ELMADJIAN, F., and PINCUS, G., Adrenal cortical responsivity of psychotic patients in relation to electroshock therapy. *Psychosomat. Med.,* **12,** 73 (1950).

MALAMUD, W., HOAGLAND, H., and KAUFMAN, I. C., A new psychiatric rating scale. *Psychosomat. Med.,* **8,** 243 (1946).

MALAMUD, W., and SANDS, S. L., A revision of the psychiatric rating scale. *Am. J. Psychiat.,* **104,** 231 (1947).

PINCUS, G., and HOAGLAND, H., Steroid excretion and the stress of flying. *J. Aviation Med.,* **14,** 173 (1943).

PINCUS, G., and ELMADJIAN, F., The lymphocyte response to heat stress in normal and psychotic subjects. *J. Clin. Endocrinol.,* **6,** 295 (1946).

PINCUS, G., Studies of the role of the adrenal cortex in the stress of human subjects. *Recent Progress in Hormone Research,* Vol. I, p. 123. Academic Press, New York, 1947.

PINCUS, G., ROMANOFF, L. P., and CARLO, J., A diurnal rhythm in the excretion of neutral reducing lipids by man and its relation to the 17-kestosteroid rhythm. *J. Clin. Endocrinol.,* **8,** 221 (1948).

PINCUS, G., HOAGLAND, H., FREEMAN, H., ELMADJIAN, F., and ROMANOFF, L. P., A study of pituitary-adrenocortical function in normal and psychotic men. *Psychosomat. Med.,* **11,** 74 (1949).

Pincus, G., Schenker, V., Elmadjian, F., and Hoagland, H., Responsivity of schizophrenic men to pituitary adrenocorticotrophin. *Psychosomat. Med.,* **11,** 146 (1949).

Pincus, G., Hoagland, H., Freeman, H., and Elmadjian, F., "Adrenal Function in Mental Disease," in *Recent Progress in Hormone Research,* Vol. IV, p. 291. Academic Press, New York, 1949.

Pincus, G., and Hoagland, H., Adrenal cortical responses to stress in normal men and in those with personality disorders. *Am. J. Psychiat.,* **106,** 651 (1950).